Design for Climate Change

Bill Gething

with Katie Puckett

RIBA Publishing

© Bill Gething and Katie Puckett, 2013

Published by RIBA Publishing,
15 Bonhill Street, London EC2P 2EA

ISBN 978 1 85946 448 9

Stock code 77532

The rights of Bill Gething and Katie Puckett to be identified as the Authors of this Work have been asserted in accordance with the Copyright, Design and Patents Act 1988.

British Library Cataloguing in Publications Data
A catalogue record for this book is available from the British Library.

Commissioning Editor: Lucy Harbor
Project Editor: Neil O'Regan
Designed and typeset by: Alex Lazarou
Printed and bound by: Butler Tanner and Dennis Ltd

While every effort has been made to check the accuracy and quality of the information given in this publication, neither the Author nor the Publisher accept any responsibility for the subsequent use of this information, for any errors or omissions that it may contain, or for any misunderstandings arising from it.

RIBA Publishing is part of RIBA Enterprises Ltd.
www.ribaenterprises.com

CONTENTS

FOREWORD

The buildings and issues described in this book were all part of the Design for Future Climate, Adapting Buildings programme, the largest programme focusing on the climate-change adaptation of buildings in the UK. This £5m programme from the Technology Strategy Board (TSB) is aimed at improving the climate resilience of building projects worth a combined capital value of £4.2bn. I am very pleased that RIBA Publishing has published this book, making the main learning points from the first half of the programme easily accessible to architects and building professionals interested in developing their own adaptation expertise.

The Technology Strategy Board is the UK's national innovation agency, sponsored by the Department for Business, Innovation and Skills and tasked with supporting research and technology development and exploitation for the benefit of UK business, in order to increase economic growth and improve quality of life. Back in 2009, we identified climate adaptation as a business opportunity which we thought the UK's world-class building-design industry could develop for the home market and exploit overseas markets.

The climate is changing and buildings need to respond to this, potentially requiring sequential interventions to keep occupants safe, comfortable and productive in an affordable way despite rising temperatures and changes in rainfall over the next century. Those responsible for buildings will want to consider options for improving asset resilience and minimizing future maintenance costs through undertaking adaptation measures as part of a building improvement regime. UK businesses can generate new income streams through providing much-needed climate-adaptation expertise.

But how should design teams develop adapted designs for climate change?

In 2009, prior to the Design for Future Climate, Adapting Buildings programme, the design industry was unable to justify the initial cost of developing climate-adaptation expertise as the immediate benefits to clients were unclear. We engaged Bill Gething as an advisor to identify how best to remove this initial barrier to providing climate-adaptation services to clients. The solution was to provide funding to create climate-adaptation strategies for current new-build and refurbishment projects. By developing strategies for specific UK building projects, those involved learned how to provide adaptation services which offer overall cost benefits to clients.

A total of 26 projects were funded in 2010 and a further 24 in 2011, ranging from the masterplanning of new towns to the detailed design of new commercial developments and refurbishment of existing buildings. The buildings selected were chosen from around 150 joint client/design-team applications all of which were required, as a condition of entry, to be targeting high standards of environmental design – at least a Very Good rating under BREEAM (Building Research Establishment Environmental Assessment Methodology). Each was awarded up to £100,000 for design fees focused on improving the resilience and adaptation of these buildings to the projected future climate.

In 2010, the Technology Strategy Board commissioned the report Design for Future Climate: Opportunities for Adaptation in the Built Environment, in order to highlight climate change as a risk to buildings and to point to adaptation measures that could overcome that risk. This report was published in June 2010, and is available at www.innovateuk.org/adaptation. It is a useful resource for those approaching their first adaptation project in the context of climate change.

We are pleased that the design teams delivering these adaptation strategies include 240 organisations and companies, all of whom are learning through their projects. The financial benefits of adaptation are exploited through the process of developing a strategy. For most projects where the client was engaged in developing the adaptation strategy, some recommendations to adapt the building to be climate-change resilient were adopted.

These projects highlight the challenges and opportunities of adapting to climate change, and they demonstrate exciting approaches, which many clients will see the benefit of investing in.

There is a growing opportunity for design teams to improve the comfort and resilience of our building stock over the next century, and I hope this book is useful in sharing the best current knowledge on adaptation with interested design professionals.

DR FIONNUALA COSTELLO
Lead Technologist, Low Impact Buildings
Technology Strategy Board

PREFACE

The 2010 report, Design for Future Climate: Opportunities for Adaptation in the Built Environment, was intended as a primer to support teams involved in the Technology Strategy Board's Design for Future Climate programme. It provided background information and some pointers as to the issues that a changing climate may raise for the construction industry and its clients. Rather than fund theoretical research, the Technology Strategy Board (TSB) chose to provide substantial additional design fees to live projects in order to enable the teams and their clients to thoroughly explore the potential impacts of a changing climate on their proposals for real buildings.

This book is intended to capitalise on the work carried out by the first tranche of projects, now substantially complete, and to draw lessons from the enormous amount of information that they have since generated. We had initially intended to focus on a strictly limited number of projects, but it quickly became clear that so much good and useful research had been carried out across the board that to restrict our scope would risk ignoring much that could be of value to the industry. We therefore decided to cover as much ground as possible, reading project reports as they became available and speaking to many of the individuals involved. Inevitably, in attempting to pull together common threads and examples of how teams tackled various issues, we may have missed some gems – either through the sheer weight of material to be reviewed or because teams had not reached a point at which they could report findings. We must apologise if we have not been able to capture their latest thinking. However, all the project reports are available through the TSB_connect website[1] for those wishing to obtain further detail or to explore particularly relevant examples. Selected case studies are also available from the Chartered Institution of Building Services Engineers (CIBSE), tailored to an engineering audience and are indicated in the project list in Appendix 1.

The map overleaf shows the geographical spread of projects and the building types involved.

We did not attempt to delve into the outputs of the second tranche of projects, but it will be interesting to see how methodologies and strategies continue to develop – particularly among those teams who also participated in the first tranche.

We would like to thank the many people who took time to speak to us about their work in such detail, to advise us on technical issues or to read and comment on our drafts. Needless to say, all of the mistakes here are our own.

Particular thanks must go to Fionnuala Costello and Mark Wray at the TSB for their support and patience; Gerry Metcalfe of UKCIP for his guidance and perceptive review of our drafts; Matthew Eames and Tristan Kershaw of the Centre for Energy and the Environment at the University of Exeter; and Tim Reeder at the Environment Agency, Adam Ritchie at Max Fordham, Rajat Gupta at the Low Carbon Building Research Group of Oxford Brookes University and Julian King at AECOM, all of whom commented on drafts of individual sections. We would also like

Project locations

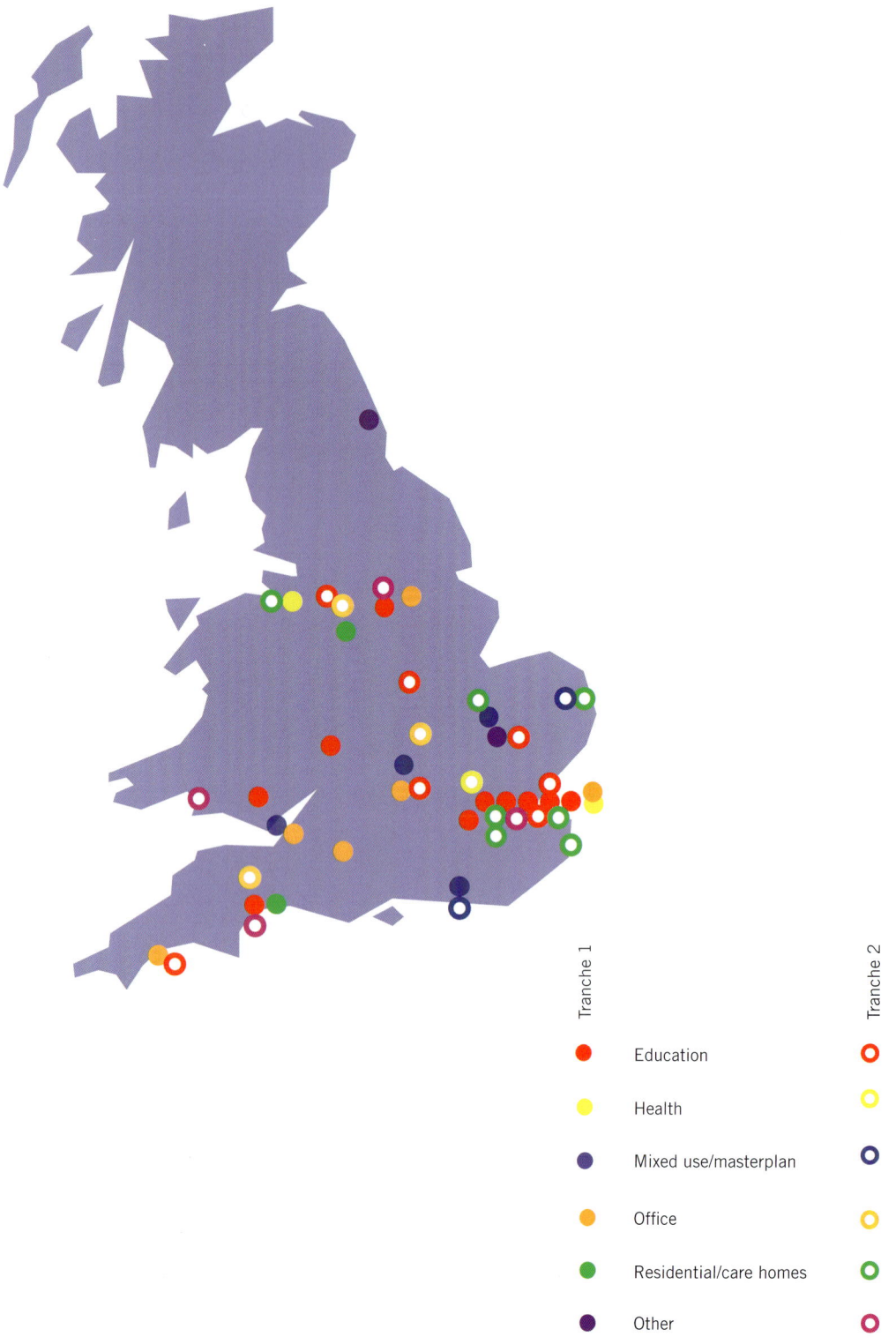

Tranche 1   Tranche 2

- ● Education   ○ Education
- ● Health   ○ Health
- ● Mixed use/masterplan   ○ Mixed use/masterplan
- ● Office   ○ Office
- ● Residential/care homes   ○ Residential/care homes
- ● Other   ○ Other

Design for Future Climate – tranches 1 and 2

to thank Anastasia Mylona of CIBSE, Maria Shamash of UKCIP, Tessa Barraclough of Peabody, Polly Turton and Jake Hacker of Arup, and sustainability consultant Brian Spires for their specialist input and, of course, all of the people involved in the projects to whom we spoke in the course of our work.

The TSB should be applauded for providing the industry with an opportunity to explore the issues in such depth and for enabling design teams to road-test solutions on live projects, in the context of real practice. When designers are excited and intrigued by a problem, their enthusiasm usually means they go much further than strictly contracted. This has clearly been the case for many of the projects, and this, coupled with the realities of the project context, has brought the issues into sharp focus and yielded worthwhile conclusions that would not have been possible in a dry academic study.

The construction industry is often criticised for not investing in research and development, while academics are often regarded as working in isolation with only fleeting contact with unpaid "industry partners". This TSB project demonstrates the potential for industry and academia to work constructively and productively together.

We hope that we have been able to do justice to such a valuable research programme.

BILL GETHING & KATIE PUCKETT

1 AGENDA FOR A CHANGING CLIMATE

Warming of the climate system is unequivocal, as is now evident from observations of increases in global average air and ocean temperatures, widespread melting of snow and ice and rising global average sea level.

Fourth Assessment Report from the UN Intergovernmental
Panel on Climate Change (IPCC), 2007

There is an overwhelming scientific consensus that:

- the climate is changing
- these changes are very likely due to increased global greenhouse-gas concentrations resulting from human activity, particularly from the use of fossil fuels
- these changes will continue if we remain on our current path, with increasingly severe consequences for all life on the planet
- there is considerable momentum in the climate system; even if greenhouse-gas concentrations were to be stabilised, warming and sea-level rise would continue for decades.

Given the scale of observed change, considerable scientific effort has been made to understand the global climate system and the reasons behind the warming, and to predict how the climate might change in future. In 1979, the first World Climate Conference, organised by the World Meteorological Organisation (WMO), expressed concern that "continued expansion of man's activities on earth may cause significant extended regional and even global changes of climate", and called for global cooperation to explore future changes and their impact on human development.

As a result, in 1988 the Intergovernmental Panel on Climate Change (IPCC) was established by the United Nations Environment Programme and the WMO to assess the risk of human-induced climate change, its potential impacts and the options for adaptation and mitigation.

From 1990, the IPCC has issued a series of reports based on peer-reviewed and published research, which have tended to grow in detail, evidence and confidence. The latest is the 2007 Fourth Assessment Report, quoted above, which reported findings based on 23 sophisticated climate models from 11 countries around the globe, including two from the UK's Met Office Hadley Centre, one of the world's leading centres for climate-science research. The report included the graph below, which indicates the potential extent of global warming.

The graph plots temperatures through the 21st century for three greenhouse-gas-emission scenarios, as well as a plot showing what would happen if greenhouse-gas concentrations were held at year-2000 values. The three future scenarios were selected from six marker scenarios used in the Fourth Assessment Report, to which the bars on the right-hand side of the graph refer. The bars show the "likely" spread of values in 2100 for each marker scenario, indicating the range of uncertainty associated with the modelling results. The scenarios and the issue of uncertainty are described in more detail in Chapter 2.

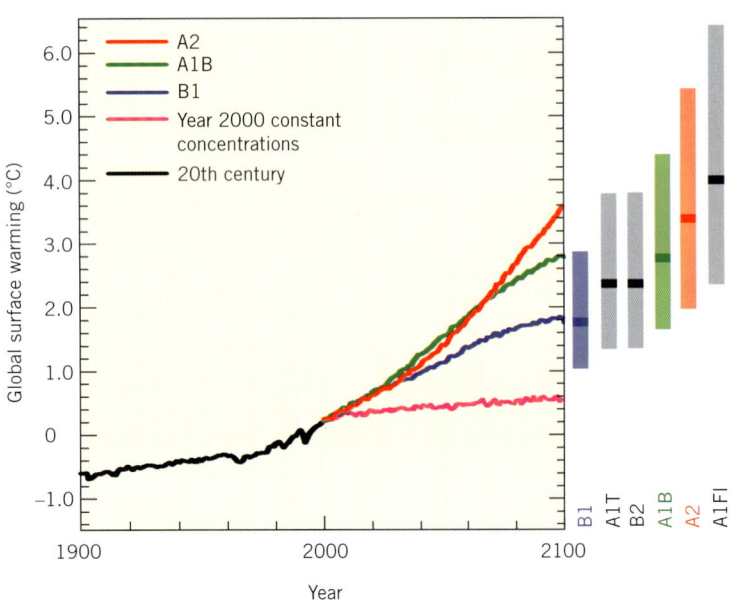

1.1 The graph above is taken from the IPCC's Fourth Assessment Report on Climate Change (2007). It shows the global averages of surface warming (relative to 1980-99) for three different greenhouse-gas emissions scenarios, as well as a theoretical scenario where emissions are held at year-2000 values. To the side of the graph, the bars indicate the best estimate (solid line) and likely range of temperature change by 2090-99 for six emissions scenarios. Emissions scenarios are described in more detail in the next chapter.
Source: Climate Change 2007: Synthesis Report. Contribution of Working Groups I, II and III to the Fourth Assessment Report of the Intergovernmental Panel on Climate Change, figure 3.2 (left panel). IPCC, Geneva, Switzerland

The graph highlights two points:

- the climate has already changed measurably since the baseline of 1990. We are not starting from zero
- up to 2050, there is little real difference between the scenarios; however, beyond the 2050s, they diverge, demonstrating the paramount importance of reducing emissions.

It is worth noting that while climate models vary in their predictions of the speed and magnitude of warming, there are no credible models that show global temperatures remaining steady or decreasing.

The report also sets out the potential impact of rising temperatures on the planet's ecosystems, human settlements and way of life, summarised in the table overleaf from the UK Treasury's 2006 Stern Review. There is a high level of confidence that a rise of more than 2°C would result in very significant impacts for all aspects of life, particularly coastal flooding and species extinction. Coastal flooding on this scale is not about the loss of a few distant islands or remote wetlands – their trading origins mean that most of the world's great commercial centres are in coastal locations. By 2050, the UN estimates that 70% of the world's population will live in cities, which means that it will be heavily concentrated in delta areas.

The potential impacts of climate change have set the agenda for successive United Nations Climate Change conferences, which have led to a general acceptance that warming must be limited to a 2°C rise if we are to avoid catastrophic impacts. This is the stark criterion on which ongoing attempts to reach international agreement to limit global emissions are based.

The issues are thus clear, urgent and interlinked. We need both to reduce the emissions that drive climate change (mitigation) and to deal with the physical effects of inevitable changes that are already under way (adaptation). The more successful we are at the former, the less will be the need for the latter.

For the construction industry, this means pursuing the now-familiar low-energy low-carbon design agenda that is increasingly embedded in legislation, but also recognising that we need to design differently in order to adapt to changes such as higher summer temperatures and to cope with more extreme events.

The UK government has led the way internationally in developing policies to tackle the dangers of climate change. In 2008, the Climate Change Act was passed, setting the world's first long-term legally binding targets for emissions reduction (80% by 2050) and establishing the independent Committee on Climate Change. Its remit is to advise the government on setting and meeting carbon budgets and monitoring its progress, and also informing and monitoring progress on adaptation through the Adaptation Sub-Committee.

Temp rise (°C)	Water	Food	Health	Land	Environment	Abrupt and Large-Scale Impacts
1°C	Small glaciers in the Andes disappear completely, threatening water supplies for 50 million people	Modest increases in cereal yields in temperate regions	At least 300,000 people each year die from climate-related diseases (predominantly diarrhoea, malaria, and malnutrition) Reduction in winter mortality in higher latitudes (Northern Europe, USA)	Permafrost thawing damages buildings and roads in parts of Canada and Russia	At least 10% of land species facing extinction (according to one estimate) 80% bleaching of coral reefs, including Great Barrier Reef	Atlantic Thermohaline Circulation starts to weaken
2°C	Potentially 20 - 30% decrease in water availability in some vulnerable regions, e.g. Southern Africa and Mediterranean	Sharp declines in crop yield in tropical regions (5 - 10% in Africa)	40 – 60 million more people exposed to malaria in Africa	Up to 10 million more people affected by coastal flooding each year	15 – 40% of species facing extinction (according to one estimate) High risk of extinction of Arctic species, including polar bear and caribou	Potential for Greenland ice sheet to begin melting irreversibly, accelerating sea level rise and committing world to an eventual 7 m sea level rise
3°C	In Southern Europe, serious droughts occur once every 10 years 1 - 4 billion more people suffer water shortages, while 1 – 5 billion gain water, which may increase flood risk	150 - 550 additional millions at risk of hunger (if carbon fertilisation weak) Agricultural yields in higher latitudes likely to peak	1 – 3 million more people die from malnutrition (if carbon fertilisation weak)	1 – 170 million more people affected by coastal flooding each year	20 – 50% of species facing extinction (according to one estimate), including 25 – 60% mammals, 30 – 40% birds and 15 – 70% butterflies in South Africa Collapse of Amazon rainforest (according to some models)	Rising risk of abrupt changes to atmospheric circulations, e.g. the monsoon Rising risk of collapse of West Antarctic Ice Sheet Rising risk of collapse of Atlantic Thermohaline Circulation
4°C	Potentially 30 – 50% decrease in water availability in Southern Africa and Mediterranean	Agricultural yields decline by 15 – 35% in Africa, and entire regions out of production (e.g. parts of Australia)	Up to 80 million more people exposed to malaria in Africa	7 – 300 million more people affected by coastal flooding each year	Loss of around half Arctic tundra Around half of all the world's nature reserves cannot fulfill objectives	
5°C	Possible disappearance of large glaciers in Himalayas, affecting one-quarter of China's population and hundreds of millions in India	Continued increase in ocean acidity seriously disrupting marine ecosystems and possibly fish stocks		Sea level rise threatens small islands, low-lying coastal areas (Florida) and major world cities such as New York, London, and Tokyo		
More than 5°C	The latest science suggests that the Earth's average temperature will rise by even more than 5 or 6°C if emissions continue to grow and positive feedbacks amplify the warming effect of greenhouse gases (e.g. release of carbon dioxide from soils or methane from permafrost). This level of global temperature rise would be equivalent to the amount of warming that occurred between the last age and today – and is likely to lead to major disruption and large-scale movement of population. Such "socially contingent" effects could be catastrophic, but are currently very hard to capture with current models as temperatures would be so far outside human experience.					

Note: This table shows illustrative impacts at different degrees of warming. Some of the uncertainty is captured in the ranges shown, but there will be additional uncertainties about the exact size of impacts (more detail in Box 3.2). Temperatures represent increases relative to pre-industrial levels. At each temperature, the impacts are expressed for a 1°C band around the central temperature, e.g. 1°C represents the range 0.5 – 1.5°C etc. Numbers of people affected at different temperatures assume population and GDP scenarios for the 2080s from the Intergovernmental Panel on Climate Change (IPCC). Figures generally assume adaptation at the level of an individual or firm, but not economy-wide adaptations due to policy intervention (covered in Part V).

1.2 This table is taken from the Stern Review, produced for the UK Treasury (2006). It shows a range of disastrous impacts if global temperatures rise by more than 2°C.

Source: The Stern Review Report © Crown copyright 2006, The Economics of Climate Change: The Stern Review

© Cambridge University Press 2007

The government's National Adaptation Programme is led by the Department for Environment, Food and Rural Affairs (DEFRA). Its first step was to produce a UK Climate Change Risk Assessment (CCRA) in January 2012, to be updated every five years. The key risks identified for buildings were:

- damage due to flooding and coastal erosion
- overheating
- increasing impact from the urban heat island effect (see Chapter 2)
- subsidence.

Other risks directly relevant to the built environment included:

- water supply shortage
- increased water demand for energy generation
- higher energy demand for cooling
- flood risk to energy infrastructure
- heat damage/disruption to energy infrastructure.

Government departments have also been required to draw up Departmental Adaptation Plans, setting out key climate-change risks and priorities. These were published in March 2010, and updated in May 2011.

It is interesting to note an apparent disconnection between our mitigation policies and our approach to adaptation. The Committee on Climate Change has explored a number of ways of achieving our mitigation goal, set, as described above, in order to limit warming to below 2°C. To give an indication of the magnitude of the action required, a typical reduction pathway would be to cut emissions by 4% year-on-year, starting in 2016. This scenario is plotted overleaf, against the Low (B1), Medium (A1B) and High (A1FI) emissions scenarios used by the Met Office Hadley Centre for UK climate projections.

The graph shows all too clearly that the three scenarios used as the basis for adaptation take a more pessimistic, or perhaps realistic, view of future emissions reduction. Even the Low emissions scenario represents a failure of our global mitigation goal and, by implication, of our attempt to avert catastrophic change.

It is all too easy for the challenge of adaptation to become an intriguing academic puzzle – for designers to become mesmerised by the interaction between projection data and theoretical building models without stepping back to consider the implications for the social and economic context within which their buildings will exist. The teams involved in the Design for Future Climate programme can be forgiven for not attempting to address this wider context, but policy makers must think more broadly. The construction industry may be able to deliver buildings that would in theory perform beautifully in temperatures even 4°C above current levels – but what will the

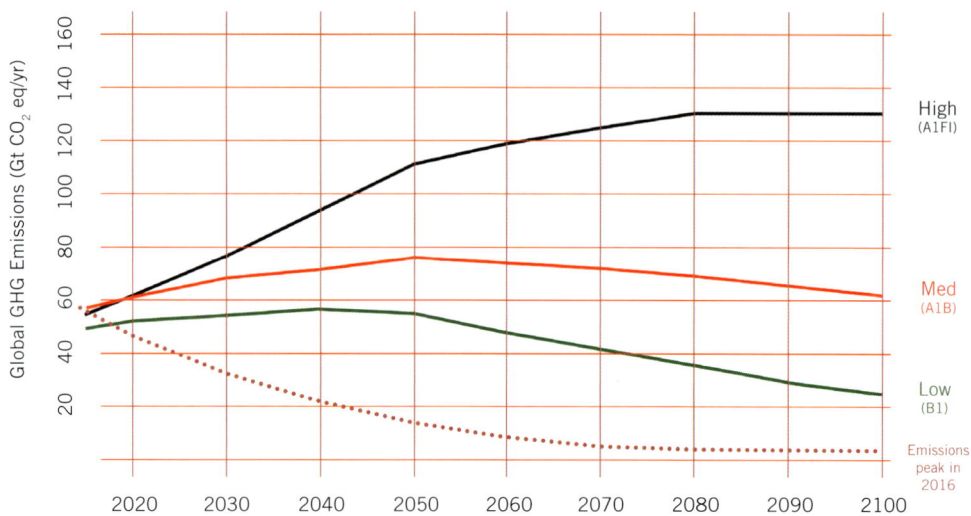

1.3 One possible scenario for avoiding catastrophic climate change is for global emissions to peak in 2016 and continue to decline by 4% each year after that. This graph shows how this compares to the high, medium and low emissions scenarios used to inform climate change adaptation strategies.

world around them be like? What will be the human and economic consequences of this level of change in locations that are not blessed with our benign climate? And how will this affect "normal" life in the UK?

● ● ●
CLIMATE CHANGE IN THE UK

Trends observed in the UK reflect global patterns. The Central England Temperature (CET) record – the oldest-established instrumental record of temperature in the world – shows that after a period of relative stability for most of the 20th century, our annual average temperature has risen by about 1°C since the 1970s.[2]

Observed trends are projected to continue and can be summarised as follows:

- warmer, wetter winters
- hotter, drier summers
- rising sea levels
- increased extreme weather events.

Weather events that are currently regarded as extreme are useful illustrations of what is projected to be normal in future. For example, the exceptionally hot summer of 2003 is likely to become the norm by the second half of this century.

DESIGN AGENDA

The impact of climate change is particularly pertinent to the construction industry, simply because buildings last a long time. Our existing built environment, and every aspect of how we live our lives, has evolved in response to a particular climate. Now that climate is changing, and may soon be significantly different. We face a real challenge in converting and upgrading our urban fabric to function in a climate for which it was not designed.

This book follows the same structure as the Design for Future Climate report, published in 2010 to provide the background to inform teams embarking on their projects, in which climate-change impacts for the built environment were split into three broad categories:

- comfort and energy performance – warmer winters may reduce the need for heating, but it will be difficult to keep cool in summer without increasing energy use and carbon emissions
- construction – resistance to extreme conditions, detailing and the behaviour of materials
- managing water – both too much (flooding) and too little (shortages and soil movement).

Over recent years, the construction industry has responded to the mitigation agenda but it has not yet got to grips with adaptation. There must now be a recognition that some changes to our climate cannot be avoided, and that these changes will have a significant impact on how our buildings perform, where they should be built and how they are constructed. It should also be remembered that change is likely to be ongoing – the rate and extent being dependent on the success of global mitigation strategies – resulting in more significant change in the second half of the century.

We need to rethink the way we design, construct, upgrade and occupy buildings to accommodate this, developing approaches that are based not on past experience but on calculated projections of future climate.

CONTEXT

Climate change is a "moving target", and the challenges will differ according to location and building type. There are no universal solutions. Adaptation strategies must be thoroughly grounded in a building's context, which means not only projected changes in the climate but changes to geography, urban fabric, energy, construction and society too.

Location

The regional variations in climate that already exist across the UK (and which are sometimes ignored by building legislation and policy) are projected to increase somewhat. This could potentially exaggerate the differences between appropriate climatic design responses across the country, so evident in our vernacular tradition.

Achieving comfortable internal temperatures in the future climate in the north of the country will be easier than in the warmer urban south-east, for example. The same may be true of water resources. In more exposed areas, however, the challenge of building sufficiently robustly to deal with extreme weather events may be paramount, with materials that may have performed satisfactorily in the past being pushed beyond their capabilities.

Flooding from rivers or the sea will be an absolute focus in some locations and irrelevant in others, although all regions will need to consider the impact of extreme rainfall on roof and surface-water drainage.

Low-carbon imperative

As discussed above, the mitigation agenda is paramount, while the age of cheap fossil fuels is also drawing to a close. There is no guarantee that we will be able to replace these with similarly cheap, sufficiently large and reliable sources of low-carbon or renewable energy. Warmer winters may reduce the need for and cost of heating, but, in summer, the cost of running buildings that rely on energy-intensive mechanical cooling rather than intelligent passive design, which minimises or avoids cooling-energy consumption, is likely to be an unwelcome burden for their occupants in a future low-carbon world.

Adaptation and mitigation measures can complement each other, but there can also be direct conflicts. For example, incorporating large areas of glazing in an attempt to reduce lighting-energy use can lead to overheating in summer, and even mid-season in highly insulated buildings, unless solar gain is carefully controlled.

The existing building stock

As with the mitigation agenda, the existing stock is the real challenge: adaptation is a conversion, not a new-build, challenge.

Many older buildings, traditionally constructed in heavy materials, with small windows and good ventilation (both controlled and uncontrolled), have performed well during recent heatwaves. But many late-20th-century examples – characterised by lightweight, poorly insulated construction, large expanses of unshaded glazing and poor ventilation – already perform badly in summer and will become either unbearable or, if they rely on air conditioning, prohibitively expensive to run.

When upgrading existing buildings to improve their winter thermal performance by increasing airtightness and insulation, we need to take care that they do not then overheat in summer.

Adapting external spaces presents a parallel challenge to building design. Generally speaking, our wider urban fabric has also evolved to take advantage of limited sunshine, rather than to provide shade and protection from it.

Robustness

We expect our buildings to perform reliably with little if any maintenance. But familiar materials may behave differently under future conditions, and they may need to be substituted or fixed differently. Adaptation strategies that rely on new materials, components or approaches to construction will also need to demonstrate their long-term capabilities, and that they can be easily maintained.

Given that climate change is a moving target, even new buildings may need to be further adapted over time. There must be both an ongoing strategy for incorporating additional measures, and the physical provision to enable those adaptations to be carried out to the standard of the original build.

● ● ●
TACTICS

A changing climate poses intriguing challenges for design and construction, and it is tempting to assume that every aspect must be addressed by ingenious or innovative design, comprehensively and immediately. Designers are often involved at the inception of a building project, advising their client on feasibility, helping to set the brief and, with an emerging agenda like adaptation, making their client aware of wider issues that they may not have taken into account. Designers should keep an open mind on alternative ways of approaching the challenge, which may offer better value for money than tackling them head-on in the confines of a building's design.

Behaviour

Some problems can be circumvented simply by altering our behaviour rather than incorporating new design features or devices. Some of these are in the control of an individual client, such as not building in a potential flood zone or relaxing dress codes to compensate for higher temperatures. It might even be possible for an individual business to alter working hours to avoid those times of the day when internal conditions are most stressed, but, realistically, a wider coordinated response would be required for such measures to be effective.

Given that the existing stock presents the overwhelming adaptation challenge, much of which will be least able to cope with climate change, it seems likely that these approaches will be explored, particularly under extreme conditions – a comfort equivalent of the hosepipe ban, perhaps? Perhaps the best example of this "soft" approach is the "Cool Biz" campaign introduced by the Japanese government in summer 2005 to reduce electricity consumption by air conditioning. The set-point temperature for government offices was raised to 28°C and staff were encouraged to forgo jackets and ties and to wear short-sleeved shirts. The campaign was relaunched as "Super Cool Biz" in 2011 after the Tohoku earthquake and tsunami forced the closure of many of the country's nuclear power plants, resulting in extreme shortages of electricity. Clearly, a designer will not be able to rely on such radical communal measures under most circumstances, but they should certainly be considered by policy makers.

Timing

Climate change is a gradual and ongoing process, subject to multiple layers of uncertainty. Some aspects of a building must be designed for its whole life, such as the foundations or principal structure. But others, such as glazing systems and services, have shorter life expectancies and will require replacement or maintenance at regular intervals. While it is sensible to develop a comprehensive long-term adaptation strategy for the changing climate, it also makes sense to hedge one's bets as far as possible, with a set of interventions that can be implemented as part of a refurbishment cycle when the evidence for their need becomes clearer. However, care needs to be taken not to design-in blind alleys, whereby a strategy will require very disruptive interventions, or even demolition, to make further adaptation possible.

Scale

Not all of the impacts of climate change must be resolved at the level of a single building. As a society, we need to be careful that the responsibility and costs of adaptation are sensibly allocated. For example, small-scale water-treatment plants or rainwater-harvesting systems may not be the most economic or carbon-efficient way of dealing with water shortage. Individual flood-defence measures are similarly likely to be less reliable and more expensive than a comprehensive approach.

Competition, consensus and regulation

The market drivers for the adaptation of buildings are not yet clear. The key issues are the layers of uncertainty associated with climate change, and the fact that the benefits of a successful strategy may be reaped only in a time-frame considerably longer than conventional financial planning. Benefits may also not be easily valued in financial terms, with dependent costs that are highly uncertain given the time-frame. Neither may benefits accrue to those financing the capital cost of the project, or indeed to any individual or organisation, but only to wider society.

For designers, the lack of an obvious financial case for adaptation measures may limit the marketing potential for adaptation design services. Given the current lack of consensus on which data to use for analysis, and the range of options that may therefore need to be explored, the additional design time and cost may be significant – and unattractive to clients who are, after all, under no obligation to consider the longer term. As a result, designers themselves may be reluctant to invest in research and training to develop these services.

More research is needed on a range of costs – of adaptation measures themselves, of the associated design time, and of the financial and other benefits – in order to strengthen the commercial case. The government also has a clear role in developing consensus on the range of "reasonable" parameters that should be considered, to enable adaptation to be embraced as a mainstream design issue. There may also be a role for regulation to address any shortcomings in the market in the interest of the common good.

Consideration of the impacts of future climate is new territory for both clients and the construction industry. It has the potential to radically alter the way we design, construct, use and adapt our buildings. As such, it could be a rich source of design inspiration as we develop elegant approaches to produce buildings that will be resilient in a future that is both certain (change is inevitable) and uncertain (the rate and magnitude of change is unclear), as well as meeting the challenging mitigation targets necessary to avoid catastrophic change.

2 UNDERSTANDING FUTURE CLIMATE

Climate is what you expect, weather is what you get

Robert A Heinlein, Time Enough for Love, 1974

– often quoted by meteorologists

CLIMATE PROJECTIONS FOR THE UK

There is a difference between climate and weather. Weather is what we experience from day to day, described using metrics such as temperature, precipitation, humidity, wind speed and direction, sunshine and cloud cover. Climate is the average weather experienced over a long period, typically 30 years. The averaging process means that climate data is less chaotic than weather data, and it is possible to identify trends that are difficult to discern amid the natural variability of weather patterns. However, buildings do need to be able to deal with that natural variability and, within reasonable limits of cost and likelihood, to withstand extreme conditions.

In a changing climate, we cannot just rely on historical records to make design decisions but must also take account of the likely pattern, timescale and magnitude of potential change.

Climate projections for the UK have been produced since the 1990s by the Met Office Hadley Centre. The latest set was released in 2009, and is known as UKCP09 (UK Climate Projections 2009), replacing the previous set published in 2002 (UKCIP02). This is the key source of climate information on which government departments, research organisations, insurers, and regulation and standards-setting bodies are basing their responses to climate change.

The UK Climate Impacts Programme (UKCIP) was set up by the government in 1997 to support organisations in the public, private, research and voluntary sectors to adapt to the unavoidable consequences of climate change. In 2011, the Environment Agency took over this role, but the range of tools and guidance developed by UKCIP for assessing the risks of climate change and developing adaptation strategies, some of which are described below, is still available on its website.

Emissions scenarios explained

Projections for future climate are made using increasingly sophisticated computer models, based on the interaction between greenhouse-gas emissions and the climate system. The IPCC's Special Report on Emissions Scenarios (SRES 2000) presented 40 emissions scenarios reflecting a range of demographic, social, economic and technological factors, of which three were selected for UKCP09. These differ slightly from the four that were used for the previous UKCIP02 projections, as shown on the table opposite.

The UKCP09 briefing report stresses that because emissions will be governed by human choices, relative likelihoods cannot be assigned to different scenarios. However, it is salutary to note that global emissions have continued to rise in line with the upper range of the selected scenarios, checked only slightly by the effects of recession.

Introduction to UKCP09

The headline trends identified by earlier projections are unchanged – warmer, wetter winters; hotter, drier summers; rising sea levels; more extreme events. However, the approach taken by UKCP09 better reflects inherent uncertainties. Whereas the previous projections were based on the outputs of the Hadley Centre's climate model alone, UKCP09 takes a broader view, using the outputs from a range of plausible climate models. These are weighted according to how closely they correspond with measured data when used to backcast past climate, and the resulting projections are presented as a probabilistic range rather than as single values. This new approach is statistically more robust but adds complexity. It also has the disadvantage that where there is insufficient agreement between the outputs of the models, a statistically robust, correlated projection cannot be made, resulting in a gap in the data. For example, wind data was not initially included in the probabilistic projections. Projections for wind have now been produced as a separate batch and so cannot be used in Weather Generator projections (see page 17).

UKCP09 provides projections up to 2099 for 26 atmospheric variables (listed in Appendix 2) using a 25km² grid. Customisable data is also available for any location in the UK via the UKCP09 User Interface. It includes both *changes* relative to the baseline period 1961–90 and *absolute values*, which designers need for definitive analysis.

The projections do not include any explicit representation of urban areas other than that reflected by the underlying base climate data at the resolution of a 25km² grid.

UK SCENARIO NAME ■ UKCP09 ■ UKCIP02	IPCC DESIG-NATION	KEY FEATURES	DEMOGRAPHIC	SOCIAL	ECONOMIC	TECHNOLOGICAL	ATMOSPHERIC CO_2/PPM
High ■ ■	A1FI	Economic growth Increasing equity between regions Fossil fuel energy	Global population peaks at 8.7bn by 2050 and declines towards 7bn by 2100, reflecting both low fertility and low mortality	The world becomes equally globalised. Regional disparities in per capita income gradually even out	Very rapid economic growth	Rapid introduction of new and more efficient technologies, fossil-intensive energy sources	970
Medium -high ■	A2	Economic growth Social, economic and technological development remains very fragmented	Fertility patterns across world converge only slowly, leading to continuous increase in global population to 15bn by 2100	World remains fragmented and regionally oriented, emphasis on self-reliance and preservation of local identities	Economic growth slower than in A1 and B1 storylines, fragmented development	Technological change fragmented and slower than in other storylines	856
Medium ■	A1B	Strong economic growth Increasing equity between regions Mixed energy sources	Global population peaks at 8.7bn by 2050 and declines towards 7bn by 2100, reflecting both low fertility and low mortality	The world becomes equally globalised. Regional disparities in per capita income gradually even out	Very rapid economic growth	Rapid introduction of new and more efficient technologies, a balance between fossil and non-fossil energy sources	717
Medium -low ■	B2	Sustainability Heterogeneous world Emphasis on local solutions to economic, social and environmental sustainability	Global population increases continuously, reaching 10.4bn by 2100	World remains fragmented and regionally oriented	Intermediate economic development	Technological change less rapid and more diverse than in A1 and B1 storylines	621
Low ■ ■	B1	Emphasis on global solutions to economic, social and environmental sustainability, including improved equity, without additional climate initiatives	Global population peaks at 8.7bn by 2050 and declines towards 7bn by 2100, reflecting both low fertility and low mortality	The world becomes equally globalised. Regional disparities in per capita income gradually even out	Rapid change to a service and information economy, with reductions in material-use intensity	Introduction of clean and resource-efficient technologies	549

● ● ● **2.1** This table shows the assumptions underpinning the Low, Medium and High scenarios used in the UKCP09 climate projections, as well as those for the Medium-low and Medium-high scenarios used in the earlier 2002 projections.
Source: IPCC SRES, CIBSE TM48

UKCP09 also includes information on sea-level change for the same three emissions scenarios, and for an additional extreme scenario, H++. This has been developed specifically to investigate sea-level rise and storm surges, and is regarded as high risk and low probability. It aims to reflect the effect of melting ice, which presents a major source of uncertainty in projecting sea-level rise but is not well represented in current global climate models.

● ● ● URBAN HEAT ISLANDS

Temperatures in cities are typically higher than in their rural surroundings, particularly at night. This phenomenon is called the urban heat island (UHI) effect. On the positive side, this means that buildings in cities require less winter heating than their rural counterparts, but it also exacerbates the impact of heatwaves, affecting thermal comfort and reducing opportunities to passively night-cool buildings.

The effect is typically strongest during warm weather, on clear, still nights. By the middle of the 1960s, there was an average difference of 4–6°C between the centre of London and its surroundings. More recently, extreme UHI intensities in excess of 7°C have been recorded, rising to 9°C in the August 2003 heatwave.

Urban heat islands are caused both by the concentration of heat sources such as buildings and traffic, and by the storage of solar energy in the urban fabric during the day and its subsequent release into the atmosphere at night. Urban areas also tend to be drier than the countryside, because of the lack of green space and because surface water is quickly removed by urban drainage systems. Therefore, less of the sun's energy is used in evaporation and evapotranspiration (water uptake and loss by plants), and more is available to heat the atmosphere. Conversely, daytime temperatures in urban areas may actually be lower than in their rural surroundings due to the shading effect of buildings and the absorption of solar energy.

Climate change may affect both the frequency and magnitude of extreme UHI events. AECOM are attempting to address this by developing UHI modelling tools to inform design decisions for their work on the North West Cambridge urban extension.

The effects are inherently difficult to predict, due to the complex interaction between variables such as solar radiance, cloud cover and wind, both speed and direction. It is important that theoretical models are validated against measured data, but this rarely happens. Arup, leading research for the Greater London Authority (GLA) and CIBSE into sources of future weather data for building modelling in London, examined the available data in and around the city. They found that sufficiently comprehensive data was only available for three locations (Heathrow and Gatwick airports and the London Weather Centre in Holborn). Although these examples did fairly represent rural, suburban and central urban situations, this level of data was not typically available for other cities.

Further information on the methodology and the projections themselves is available from the UKCP09 website, and a full description can be found in the UK Climate Projections: Briefing Report, sections 3 and 4.

A wide range of standard data tables, maps and graphs is available for climate variables for overlapping 30-year periods in a number of different forms. However, for those who simply need a broad understanding of the changes that we might need to deal with, UKCIP developed a series of maps specifically for the original Design for Future Climate report, which is available at www.innovateuk.org/adaptation. A sample of these is included as Appendix 3.

Understanding and using probabilistic projections

UKCP09 uses the same set of terms as the IPCC reports to describe the probability of different outcomes for each climate variable. These terms have a precision, directly linked to statistical percentages, that is in marked contrast to the hyperbole of political and media coverage of climate change:

Virtually certain	> 99% probability
Extremely likely	> 95%
Very likely	> 90%
Likely	> 66%
More likely than not	> 50%
As likely as not	50% (a central estimate)
About as likely as not	33–66%
Unlikely	< 33%
Very unlikely	< 10%
Extremely unlikely	< 5%
Exceptionally unlikely	< 1%

The three most widely used probability levels are 50% (the central estimate), and 10% and 90% (the lower and upper limits of the "likely" band of outcomes – ie it is "likely" that a variable will change by more than the 10% projection and "unlikely" that it will change by more than the 90% value).

UKCP09 also features additional tools aimed at specialist users, which allow the generation and analysis of tailored data-sets: the Weather Generator and the Threshold Detector. The Weather Generator can be used to produce synthetic, random but statistically plausible daily or hourly data at the resolution of a 5km^2 grid. Though the spatial resolution is increased, the Weather Generator does not include any additional information on climate change. Instead, data from the baseline period is used to calculate the statistical relationship between weather variables, and the UKCP09 change factors are applied to produce future probabilistic outputs. The Weather Generator should be run at least 100 times for the 30-year period in question, producing 3,000 years of data for statistical analysis. This requires specialist statistical skills and presents a heavy computational and data-handling burden, making it unsuitable as a tool for day-to-day project work. However, as discussed below, it has been used to make future weather files available in a form that can be readily used with standard building-simulation models. The Threshold Detector tool (used on the Edge Lane TIME project, see below) allows further analysis of daily data produced by the Weather Generator to show how often a user-defined parameter is exceeded.. The UKCP09 user interface provides three predefined outputs:

- Heating Degree Days (HDD) – the number of days when the mean daily temperature is below 15.5°C, and heating would be required (it should be noted that this is different to the standard definition of Heating Degree Days normally used by engineers)
- Cooling Degree Days (CDD) – the number of days when mean daily temperature exceeds 22°C
- Heatwaves – when maximum daily temperature is greater than 30°C and minimum daily temperature is greater than 15°C for a minimum of three consecutive days.

It is also possible to define custom events.

At the Edge Lane project, Oxford Brookes University used the Threshold Detector tool to show the changing demand for heating and cooling, and the incidence of heatwaves. The results are shown in the tables opposite.

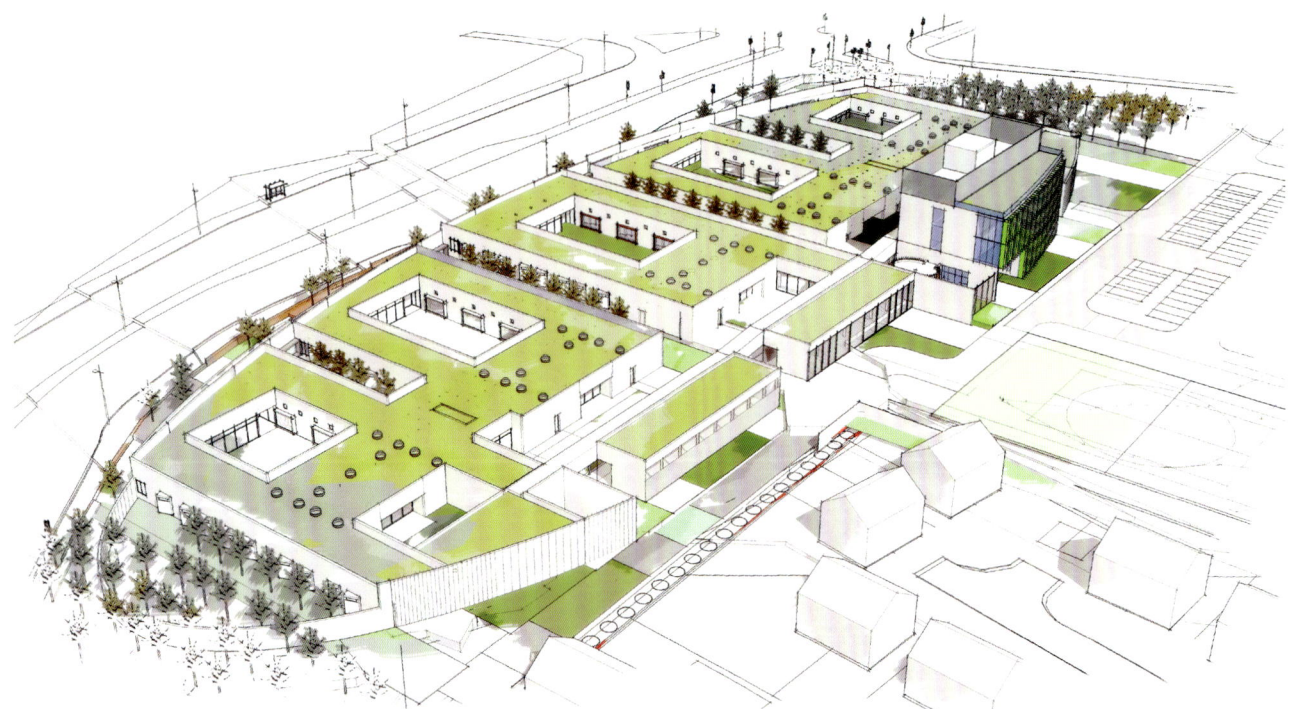

2.2 Edge Lane TIME project, a new mental health facility in Liverpool, designed by Medical Architecture.

Heating Degree Days, as projected for Liverpool by the UKCP09 Threshold Detector

HDD	JAN	FEB	MAR	APR	MAY	JUN	JUL	AUG	SEP	OCT	NOV	DEC
Baseline	31	29	31	30	29	22	13	14	24	30	30	31
2030	31	29	31	30	27	12	5	4	14	26	30	31
2050	31	28	31	29	24	9	3	3	10	24	29	31
2080	31	28	30	27	17	5	1	1	4	17	27	30

Cooling Degree Days, as projected for Liverpool by the UKCP09 Threshold Detector

CDD	JAN	FEB	MAR	APR	MAY	JUN	JUL	AUG	SEP	OCT	NOV	DEC
Baseline	0	0	0	0	0	0	0	0	0	0	0	0
2030	0	0	0	0	0	1	2	2	0	0	0	0
2050	0	0	0	0	0	1	4	4	1	0	0	0
2080	0	0	0	0	0.3	3.5	10.4	9.9	3	0.6	0	0

The growing incidence of heatwaves in Liverpool, projected by the UKCP09 Threshold Detector

HEAT WAVE	JAN	FEB	MAR	APR	MAY	JUN	JUL	AUG	SEP	OCT	NOV	DEC
Baseline	0	0	0	0	0	0	0	0	0	0	0	0
2030	0	0	0	0	0	0	0.1	0.1	0	0	0	0
2050	0	0	0	0	0	0.1	2	2	0	0	0	0
2080	0	0	0	0	0	0.2	0.7	0.7	0.1	0	0	0

● ● ●   2.3 The tables above were produced by Oxford Brookes University for the Edge Lane project in Liverpool, using the UKCP09 Threshold Detector.

• • •
MOVING FROM CLIMATE TO WEATHER

For an in-depth understanding of how climate change will affect a building design, the basic climate information needs to be "translated" into forms that designers are familiar with. This is particularly the case for the analysis of comfort and energy performance, where industry standard environmental modelling tools are typically used to test design proposals.

Using these tools requires detailed future weather data-sets that are equivalent to current "standard" weather files – such as the CIBSE Test Reference Years (TRY) and Design Summer Years (DSY) (see page 24).

When the Design for Future Climate programme was initiated, the only available future weather files were produced by CIBSE with associated guidance (TM48: The use of climate change scenarios for building simulation),[3] based on UKCIP02 projections. TRYs and DSYs were provided for three time "slices" (the 2020s, 2050s and 2080s) for each emissions scenario for 14 locations. They were produced using a "morphing" methodology, developed with Arup, which adjusted the standard CIBSE weather files in line with the projections, taking the monthly average changes set out in UKCIP02 to stretch and shift them.

By the time the Design for Future Climate projects themselves were under way, alternative future weather files were starting to emerge from the ARCC research programme (Adaptation and Resilience in a Changing Climate), based on the newer UKCP09 probabilistic climate projections and using the Weather Generator as an alternative to the established morphing technique.

Different teams chose to use different data-sets, providing a useful "road test" of the available options.

Three teams elected to use the original CIBSE TM48 weather files for a variety of reasons:

- a preference for using data-sets that were directly comparable with the standard TRYs and DSYs used on projects where design work had already started
- a preference for using data-sets that were tried and tested, and which had the CIBSE industry stamp of approval, rather than acting as guinea pigs for data-sets that were only just emerging. It should be borne in mind that all of the projects were "live", with all of the concerns over design liability that this entails
- the attraction of a comparatively simple set of data, with a single projection for each emissions scenario and time slice, as opposed to the range of choice presented by the probabilistic approach.

Of the projects which used the newer files based on the results of the Weather Generator, the majority analysed in detail for this book chose those produced under the University of Exeter's PROMETHEUS project, although two used Manchester University's COPSE files. PROMETHEUS TRYs and DSYs are available to download for 50 locations, three time periods and the 10th, 33rd, 50th, 66th and 90th percentiles for two emissions scenarios (Medium and High) from the project website. Additional sets can be produced for specific locations. The COPSE methodology produces a single future TRY and DSY for each emissions scenario for a given location. COPSE also produced Design *Reference* Years (DRYs), a new concept. These are hourly weather series which can be used to assess both heating and cooling. They consist of months of near-extreme hot, high-solar or high-humidity data and months of near-extreme cold, low-solar and low-humidity data. A TRY is first used to identify the weather variable and month that is of greatest concern for a building design, and then the relevant month from the DRY can either be incorporated into the TRY or used as a standalone data-set to test a particular vulnerability.

CIBSE is planning to release an updated set of its morphed future weather files, using the UKCP09 probabilistic projections. For each location, the new TRYs and DSYs will include three time periods (2020s, 2050s, 2080s), and the 10th, 50th and 90th percentiles for the three emissions scenarios. A beta version of these new weather files was used by Arup for the 100 City Road project.

CIBSE is also in the process of developing weather files for London that take account of the urban heat island effect (see page 16).

Morphing versus the Weather Generator

The CIBSE future weather years are produced by morphing historic TRYs and DSYs to reflect climate projections. The PROMETHEUS and COPSE weather files, on the other hand, use the UKCP09 Weather Generator to produce multiple sets of artificial weather, from which TRYs and DSYs are produced. There are ongoing discussions among experts as to the strengths and weaknesses of both methods, some of which are summarised below.

- As both are based on historic observations, past relationships between weather variables are carried forward to future projections. Therefore they cannot predict changes in those relationships.
- The averaging and selection process inherent in both approaches means that extreme events are not well represented.
- As the morphing process is applied to the baseline weather file, morphed files retain the basic "shape" of the original file, making comparisons between weather files through the century clearer.
- Weather Generator-derived files are more random and, while they may show the same overall trends, direct comparison between files is less obvious.
- The morphing methodology depends on having suitable base data for any given location. There are relatively few locations (14 at present) where sufficiently detailed historic records have been kept over a long enough period. The Weather Generator, on the other hand, is able to synthesise data for a much more detailed 5km grid, but is limited by the number of variables it includes.
- The morphed files are subject to copyright charges whereas files produced using the Weather Generator are not.

CIBSE TEST REFERENCE YEARS AND DESIGN SUMMER YEARS

The Chartered Institution of Building Services Engineers (CIBSE) has produced standard weather files based on measured hourly data for the period 1983–2004 for 14 locations around the UK, which are used to test the performance of building designs.

The Test Reference Year (TRY) is used to analyse annual energy performance. It is a "typical" year composed of 12 statistically typical months joined together using a smoothing technique to make up a composite but continuous year of data.

The Design Summer Year (DSY) is used to test for summer overheating in near-extreme conditions. This is measured data for one particular year, selected as having a near-extreme summer. The selection is made by ranking 20 years based on the average dry-bulb temperature over the period from April to September. The DSY is the year in the middle of the upper quartile of the distribution – or the third-hottest summer in a 20-year period.

It should be noted that for certain locations records are not available over such a long period, and the DSY is based on a limited sample rather than representing a true long-term average. As a result, for these locations, modelling using the TRY can actually result in greater overheating than the theoretically more extreme DSY.

Also, the current methodology for selecting a DSY may not always produce a particularly stretching test of potential overheating. The year may be warm on average, but with no particularly intense hot spells that would challenge a building's ability to remain comfortable. Arup's research for the GLA and CIBSE (see page 16) also explored the potential of a new metric: Weighted Cooling Degree Hours (WCDH) to take account of both how much a threshold temperature is exceeded and how long it is exceeded for. They analysed the probability of a similarly warm year occurring again (known as a "return period" and expressed as 1-in-10-years, for example), using the UKCP09 projections to show how warmer years occur more frequently as the century continues.

On the basis of this analysis, a set of probabilistic DSYs was proposed, exhibiting different characteristics. Some feature long periods of moderate persistent warmth, while others contain an intense warm spell or spells, as shown on the graphs opposite.

It is striking that in the longer term, by the 2050s and 2080s, the return periods for several of these unusually warm historical years reduces to just one or two years – ie all years will be as warm or warmer.

It is anticipated that the full results of the research will be published as CIBSE TM49: Probabilistic Design Summer Years for London in the near future.

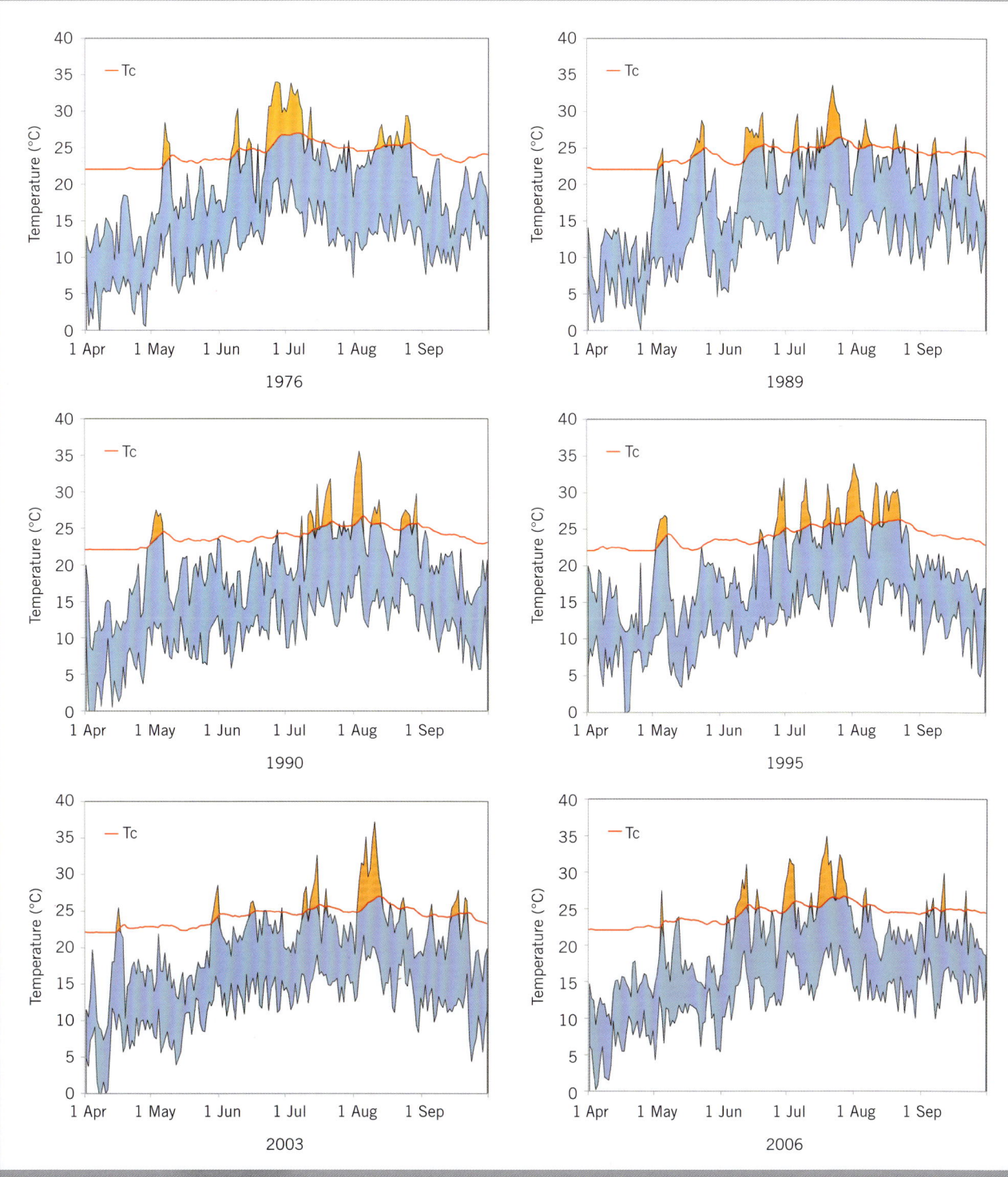

2.7 These graphs show daily minimum and maximum summer temperatures for six different Design Summer Years for London, produced by Arup for the Greater London Authority and CIBSE. The red line indicates the Adaptive Comfort Threshold (see Chapter 3 for an explanation of the ACT).

● ● ●

USING FUTURE WEATHER FILES

Designers are not generally accustomed to analysing raw climate data. It is perhaps unsurprising, therefore, that a number of teams turned to these weather files to paint a picture of general changes in external climate, even before the form of the proposed building had been developed.

For example, the Central Saint Martins team (using the earlier CIBSE future weather files) illustrated the likely increase in demand for cooling by comparing the amount of time that *external* temperatures would be above *internal* comfort thresholds for current and future London Design Summer Years. In the current London DSY, temperatures are higher than 28°C for 3.4% of the hours when a building would be occupied. Under the High emissions scenario, this reaches 30% of occupied hours by the 2080s. As summer accounts for 25% of the year, this effectively represents the whole of the summer.

The team also used the concept of Cooling Degree Days, which combines the severity and duration of outdoor temperatures. This indicated that demand for cooling could increase by a factor of between two and three by the 2080s.

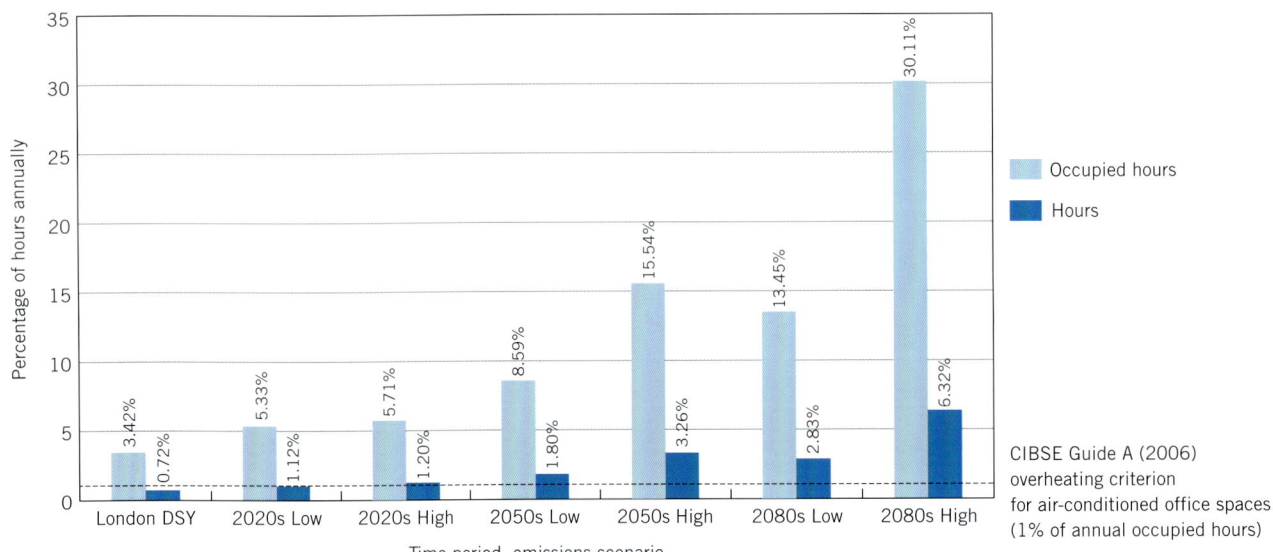

2.8 This graph was produced by the team working on the Central Saint Martins project to assess how the changing climate could make achieving internal comfort more difficult. It shows the percentage of hours annually that external temperatures will be higher than the recommended internal comfort threshold, using current and future CIBSE Design Summer Years.

2.9 Central Saint Martins University of the Arts London campus, King's Cross, designed by Stanton Williams.

2.10 100 City Road, a new commercial development in London, designed by AHMM.

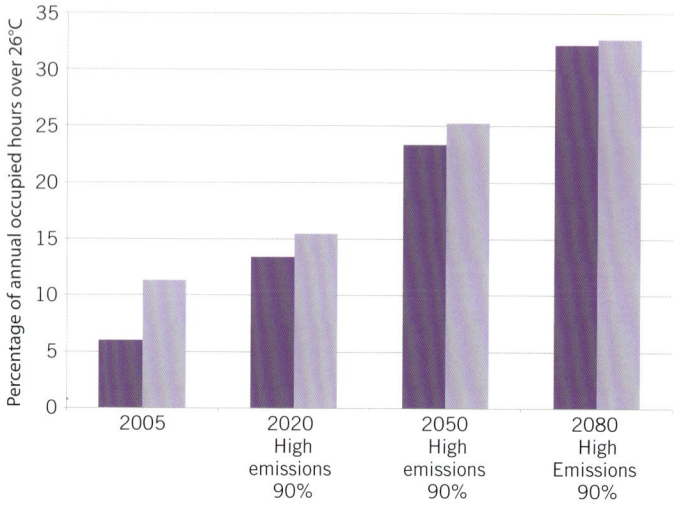

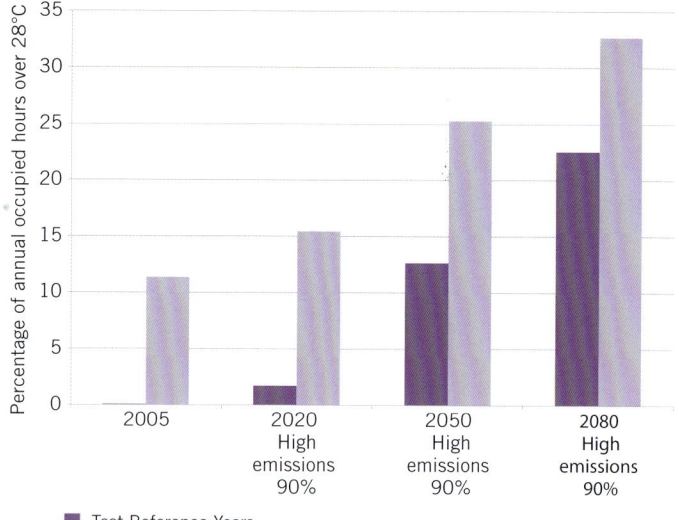

Test Reference Years
Design Summer Years

2.11 These graphs were produced by Arup for the 100 City Road project to show the increasing proportion of time when external temperatures will be higher than CIBSE benchmarks for internal comfort (26°C and 28°C), using current and future CIBSE TRY and DSY files.

At 100 City Road, Arup created the graphs above to compare external temperatures, as shown by the historic and future London weather files, with CIBSE's internal overheating criteria. By the 2080s, under the 90th-percentile High-emissions TRY, there are more than 600 hours when the external temperature is above the comfort limit of 28°C compared with fewer than 30 in the 2005 TRY. This would suggest a significant impact on the mixed-mode operation of the building – ie the use of both controlled and "free-running" modes (the latter denoting a building that does not rely on mechanical cooling systems to remain comfortable in summer).

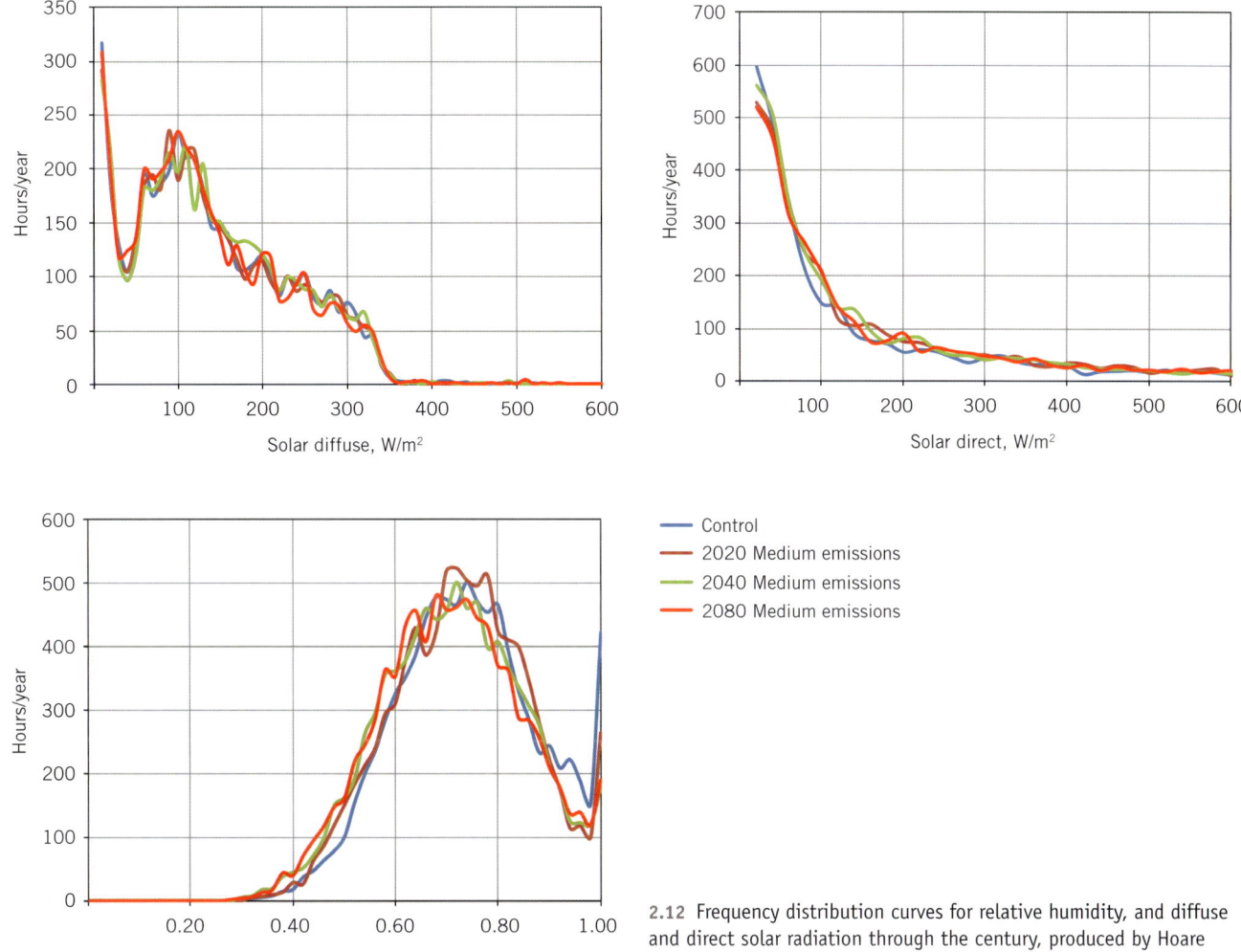

Solar diffuse, W/m²

Solar direct, W/m²

—— Control
—— 2020 Medium emissions
—— 2040 Medium emissions
—— 2080 Medium emissions

Relative humidity

2.12 Frequency distribution curves for relative humidity, and diffuse and direct solar radiation through the century, produced by Hoare Lea for the University of Greenwich project using COPSE data.

For the University of Greenwich project, Hoare Lea used frequency distribution graphs to provide a richer level of analysis than simple bar charts, showing annual exceedance of a given value. This kind of technique is necessary to reveal trends when using weather data which is inherently variable (in this case, Manchester University's COPSE data-set, produced using the Weather Generator). This variability may reflect the chaotic nature of weather, but it does mean that underlying trends can be somewhat obscured by the considerable overlap between data-sets for different time slices though the century.

This point was highlighted by Arup on the University of Sheffield Engineering Graduate School. Initial work on the project was based on the earlier, morphed, CIBSE TM48 weather files but the team switched to PROMETHEUS data when developing their alternative design. They noted that the latter files, derived using the Weather Generator, exhibit a more pronounced element of randomness than the morphed CIBSE files, although general trends were still evident when they applied data analysis techniques such as the frequency distribution curves below.

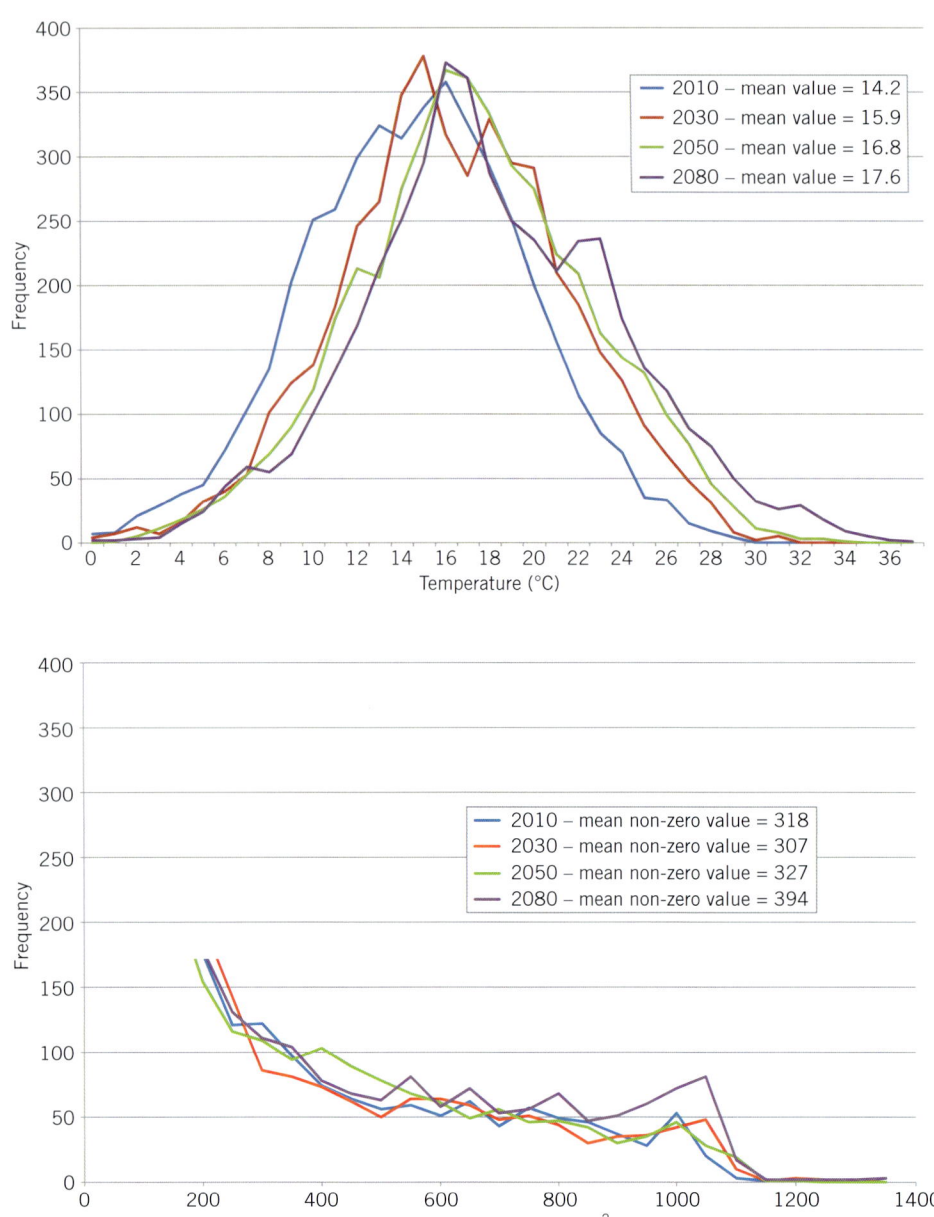

2.13 Frequency distribution curves for external dry-bulb temperature (above) and direct solar radiation (below) through the century, produced by Arup for the University of Sheffield Engineering Graduate School project using PROMETHEUS data.

The difference is very clearly shown by AEDAS' temperature plots for the Harris Academy in Purley using the the morphed CIBSE TM48 weather files. Here, temperatures are plotted for the same August day through the century. The morphing technique means that the underlying trends in the climate are obvious without the need for further analysis.

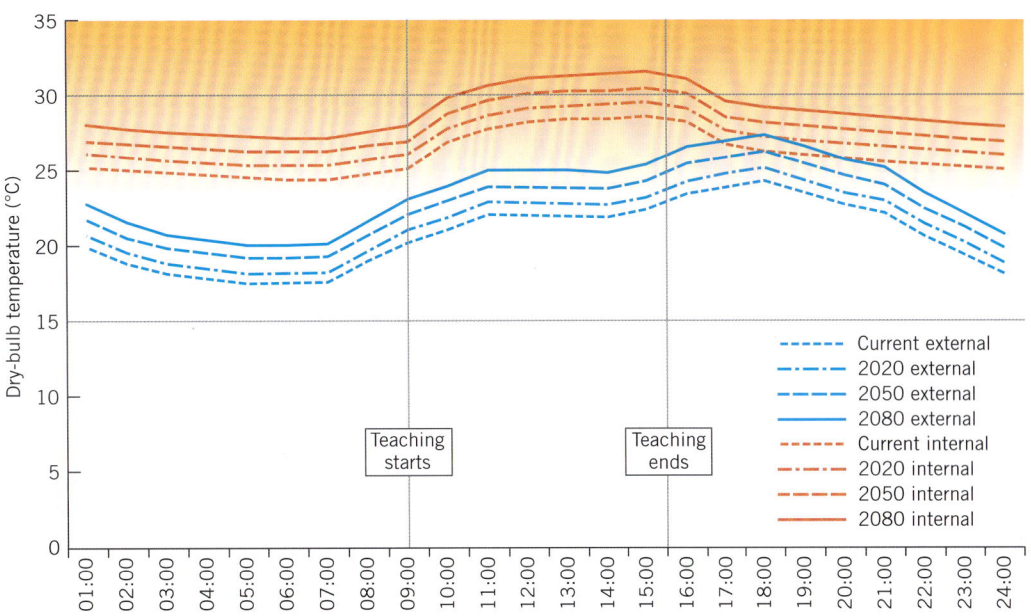

2.14 Temperature plots for internal and external temperature, produced by AEDAS for the Harris Academy in Purley using morphed CIBSE weather files.

Weather file anomalies

The teams discovered a number of anomalies and quirks in the weather files.

Arup, testing a pre-release version of morphed weather files based on UKCP09 probabilistic projections, noted that these indicated that solar gains will reduce due to increased cloud cover, which in turn would cause a rise in night-time temperature and a shift in the diurnal temperature range. This differed from Weather Generator-derived files, which indicated that solar radiation would intensify.

The University of Greenwich team also noted an increase in summer radiation, but pointed out that these apparently anomalous results produced by the Weather Generator had been raised by Napier University (a partner in developing the COPSE weather data used on the project). The relevant Weather Generator algorithm was revised and reissued in February 2011, and now shows no significant rise.

The discovery of these anomalies demonstrates the value of research projects like this, which allow designers to interrogate new data and tools in depth so that they are more robust when taken up by the mainstream. It is important that these issues are resolved fully, because they raise uncertainties in a field where there is already considerable unavoidable uncertainty.

Where there are discrepancies in derived data, it is perhaps helpful to return to the source of the projections for general design direction. Based on the UKCP09 projection maps for the Medium and High scenarios, the central estimate is that cloud cover will decrease through the century for the majority of the country. This trend will be more pronounced in the south than the north, with a slight increase in cloud cover in the far north of Scotland. The full range of likely projections shows a similar general trend towards a reduction in cloud cover, although in the 90th-percentile projections there is a slight increase first.

Climate analogues

Some teams found it helpful to use the current climate in other geographical locations as a shorthand to illustrate future characteristics of the UK climate, referred to as a "climate analogue". For example, Gale & Snowden compared the future climate of their Exeter site to that of Cologne today – a particularly useful analogy as this gave them added confidence in the use of built examples of Passivhaus design in Germany as precedents for their work.

To illustrate the most extreme projections of climate change – High emissions, 90th percentile, 2080 – Triangle Architects used Casablanca, Morocco, as an analogue for Leek, Staffordshire. Meanwhile, the Mill project team compared the projected climate for Cardiff in the 2050s to that of Porto, Portugal, today, and used Brisbane, Australia, as an analogue for the extreme scenario for the 2080s.

Analogues can be powerful in terms of illustrating potential change, but care should be taken to understand which aspects of the climate are comparable and which are not. While temperature and perhaps rainfall patterns may move closer to these distant examples, the angle of the sun and the length of day will of course remain unchanged. As a result, while some techniques for keeping cool might be applicable, shading solutions that depend on excluding the sun at particular times could not be transferred wholesale.

Different climate variables are associated with different risks, so it may be appropriate to use different weather data to assess the impacts of each. The key is to explore the sensitivity of a project to change. Is there a significant difference in the impact of a temperature rise of 0.5°C compared with a rise of 1°C, to the extent that an adaptation strategy is no longer valid? Are there particular thresholds at which a step-change in adaptation strategies is necessary, or is the transition smooth and gradual?

● ● ●
FUTURE CLIMATES EXPLORED BY THE PROJECT TEAMS

The Design for Future Climate teams reached different conclusions as to the most appropriate data to use for assessing the risks to their particular scheme. The diagram below shows the range of climates that were explored across the first tranche of projects, overlaid on a ProCliP graph for summer mean daily maximum temperature. This is a useful way of indicating which weather files were used, and the spread of the scenarios they selected, with the probabilistic range investigated in each case.

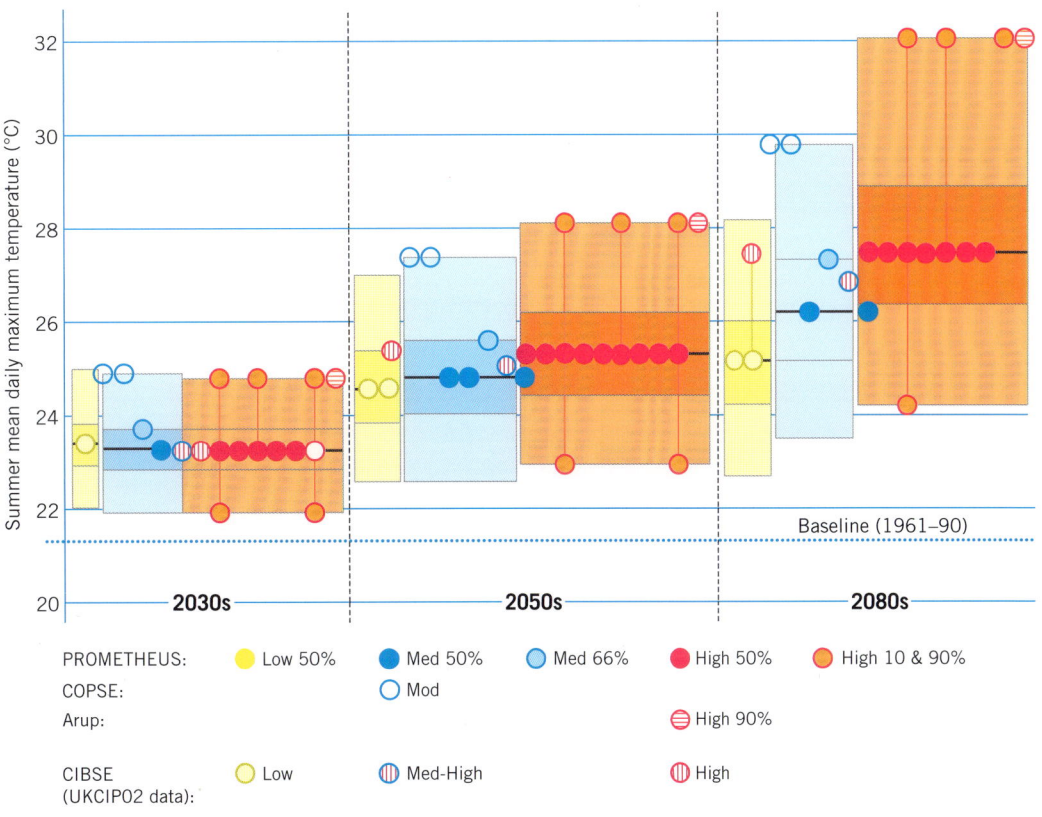

2.19 This diagram shows the weather files chosen by all of the Design for Future Climate project teams, overlaid on a ProCliP file to demonstrate the range of future climates explored by the teams.

The principle of the probabilistic climate projections is that users of the data can make informed decisions about which data to use based on the time-frame for their project and the risks to which it is most vulnerable. For the construction sector, this puts the onus on individual clients and their designers to weigh up uncertainties with which they are not familiar. Of course, it could be argued that the uncertainties associated with other design and financial decisions, routinely handled, are much greater. For example, the effect of universal IT and mobile technology on how offices are used has been more profound than anyone registered when the first PCs entered our homes and offices some 30 years ago – the time-span that we now might consider in terms of design for adaptation.

There will be many, perhaps more modest, projects where clients and designers will feel overwhelmed by the data and would be glad for the decision to be taken out of their hands, either by regulation or by guidance that narrows the choice to a manageable level.

The decisions made by the Design for Future Climate projects may be useful in helping to build some consensus on what a "reasonable" approach should be, although this might then be tempered by the circumstances of an individual project.

3 DESIGNING FOR THERMAL COMFORT: Comfort & Overheating

○ ○ ○

● ● ● As we noted in Chapter 1, our built environment and our lifestyles have evolved together in response to a specific climate. Now that climate is changing, we face the challenge of rapidly revising our accepted approach to building design to reflect a new and developing set of conditions.

All the Design for Future Climate project teams paid significant attention to thermal comfort, illustrating that of all the projected consequences of climate change, hotter summers are seen to be the most pressing issue for designers and are likely to have the most demonstrable impact on the way we design our buildings.

Until recently, the principal focus of environmental design and regulation in the UK has been on reducing heating-energy use and, latterly, the associated carbon emissions as an important part of the UK's carbon-reduction strategy. One of the positive aspects of climate change is that less energy will be needed to heat our buildings as winters become warmer. The consequent financial and carbon savings are likely to be significant, although the change will be a gradual one and any radical departure from current insulation standards and the design criteria for heating systems will not be advisable for some years. Design for winter thermal comfort must reflect today's climate, taking account of the natural variability of the weather and including the potential for extreme events in future.

Summertime thermal comfort has traditionally been less of a concern, with an expectation that most buildings should be able to provide reasonable internal conditions without mechanical assistance, relying largely on natural ventilation. Indeed, a (mature) architect who trained in the 1970s would have found not a single reference to cooling in all 300 pages of Burberry's

Environment and Services,[4] the standard textbook on environmental and building services at that time. Even CIBSE's Design Summer Year weather files, used to test overheating of free-running buildings, were only released in 2002, and initially for just three locations. (See chapter 2 for more discussion on these.) Traditional buildings – poorly insulated but constructed in heavyweight materials with relatively small windows and high levels of controlled and uncontrolled ventilation – may require a lot of heating in winter but stay reasonably cool in summer, and have generally coped quite well with occasional hot spells.

The same cannot be said of many more recently constructed buildings. Lightweight, over-glazed and poorly insulated, they often perform badly in both seasons, losing heat excessively in winter, heating up rapidly in summer and relying on mechanical systems to keep them habitable. They are already prone to overheating in a typical summer, let alone under the more extreme conditions experienced in recent years. The trend for buildings to be used more intensively, both in terms of occupant numbers and the equipment used, exacerbates the problem – as will higher insulation and airtightness standards, intended to save energy in winter, because heat can no longer escape through a leaky building fabric. The increasing density of our urban environment also adds to the challenge of keeping buildings cool, as the urban heat island effect is intensified (see page 16).

While the pressures on our buildings are increasing, so are our expectations. Comfort cooling is frequently presented as a mark of "quality" by commercial property agents, and there is a danger that ever-more-pervasive air conditioning will acclimatise us to artificially cooled spaces, to the extent that these become the norm.

Extreme events can tip the balance from one norm to another. Until recent years, for example, homes in Athens tended to have a single air-conditioning unit in one room. After the extreme hot summer of 2007, it became the norm to install a unit in every habitable room. In the UK's next extreme summer, will people opt for the quick fix of buying an air-conditioning unit for instant relief or make the effort to carry out more fundamental alterations to their homes, some of which might require planning permission? A proactive approach to design for summer comfort, exploiting passive techniques as far as possible, is essential if energy-hungry knee-jerk reactions are to be avoided.

The Design for Future Climate projects can be regarded as pioneers in this process, investigating a variety of design responses to the challenge across a wide range of building types and uses, exploiting both active and passive approaches.

It was striking that teams working on pre-existing building designs, when prompted to re-examine these from the perspective of summertime performance, were quite self-critical in identifying potential shortcomings. The process revealed how poorly many of the buildings we design will perform in summer in the current climate, let alone under more testing conditions.

The teams also identified how quickly projects that comply with current overheating criteria will start to fail, even as relatively small climatic changes take place.

In establishing how to take account of warmer summers, it is important to first understand what makes a comfortable environment and how overheating can be defined.

Designing for comfort

This table summarises some interrelationships between anticipated changes in climate and opportunities for design, and indicates the timescales to consider when developing design strategies.[†]

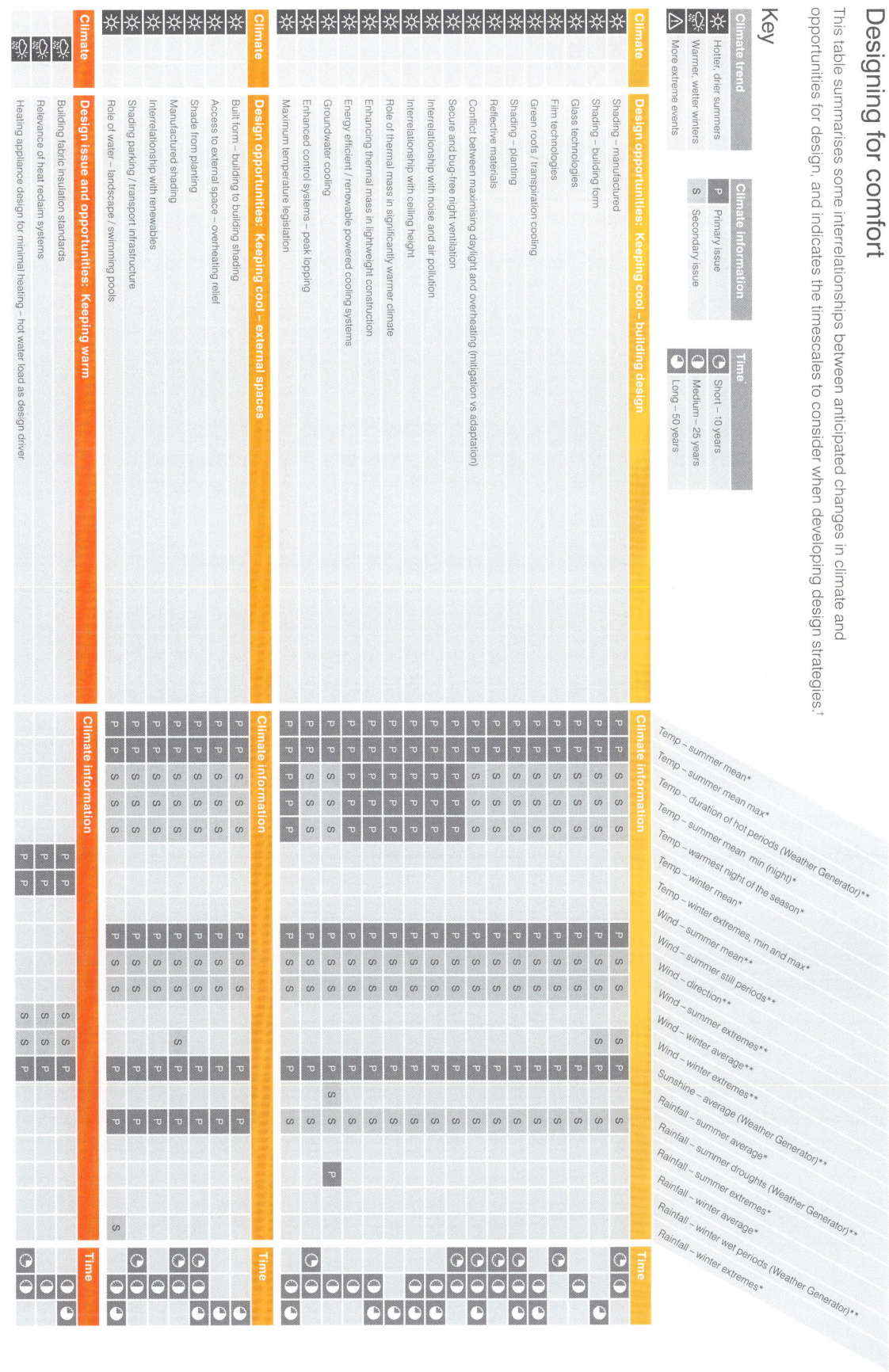

3.1 This table, taken from the original TSB report, summarises the relationships between aspects of climate change relating to thermal comfort and opportunities for designers, and gives an indication of the necessary timescale for strategies. Reprinted from Design for Future Climate – Opportunities for adaptation in the built environment, TSB, 2010.

WHAT IS THERMAL COMFORT?

Thermal comfort is inherently a personal and subjective concept. The British, European and international standard BS EN ISO 7730 Ergonomics of the Thermal Environment defines it as "that condition of mind which expresses satisfaction with the thermal environment".

To be comfortable, humans must be able to maintain an acceptable balance between the heat generated by their bodies and its flow to the surrounding environment. When the balance is right, we feel comfortable. When the balance is wrong, we start to feel discomfort, though we do not suffer medical symptoms beyond irritability, sweating or shivering. However, in circumstances of extreme heat or cold, our core body temperature (37°C) can be affected, causing more serious thermal stress. A change of only a degree or two can result in hypothermia (a drop in temperature) or hyperthermia (a rise in temperature), both of which can be harmful and potentially fatal.

Comfort therefore depends on the following combination of personal and environmental factors, as outlined in CIBSE Guide A (2006) on environmental design.

Metabolic rate

The amount of heat we produce depends on our level of activity: from about 75W when we are sleeping, to just over 100W for office work to about 450W for particularly energetic dancing.[5]

Clothing

The insulation level of clothing is typically expressed as a "Clo" rating, zero being the rating for a naked person and one for a person wearing a typical business suit.[6]

Temperature

The temperature of an environment has two components: the air temperature and the radiant temperature (caused by solar gain through an unshaded window, for example). Their combined effect can be expressed as a single metric, the "operative temperature", a weighted mean used in UK, European and international standards. This replaces a similar metric, the "dry resultant temperature" used by CIBSE Guide A (2006). For low air velocities (0.1 m/s), both metrics produce similar values.

Humidity

One way that the body regulates its internal temperature is by sweating, enabling it to lose heat through evaporation. High humidity combined with higher air temperatures hampers this cooling effect.

Air movement

Some air movement is necessary to avoid feelings of stuffiness and staleness, and it also aids cooling through both convection and evaporation. Perceptions of air movement depend on the interplay with other factors – what is considered a pleasant cooling breeze in the summer can be an annoying draught in winter.

Other factors

Post-occupancy evaluation has shown that buildings that provide opportunities for people to make changes to their environment ("adaptive opportunities") are more likely to be positively rated, while those that do not are more liable to overheat. However, those who are ill, very old or very young are more vulnerable to temperature extremes – as are those who are less able to balance their internal temperature by adjusting their clothing or activity level, such as young children, hospital patients or healthy office workers constrained by working patterns or dress codes. The speed at which temperatures increase, and the duration of a peak, will also affect people's ability to cope. What you are used to plays a role too, from the regional climate in which you live and weather conditions in recent days, to whether you've just stepped out of an air-conditioned car. Regional variations are illustrated by the different temperatures used by the Met Office for their National Severe Weather Warning Service, a 4°C range in daytime temperatures across different parts of the country.

REGION	DAY °C	NIGHT °C
London	32	18
South East	31	16
South West	30	15
Eastern	30	15
West Midlands	30	15
East Midlands	30	15
North West	30	15
Yorkshire and Humber	29	15
North East	28	15

● ● ● **3.2** Temperature thresholds used by the Met Office for its National Severe Weather Warning Service.
Source: NHS Heatwave Plan for England 2012, p40

DEFINING OVERHEATING

Human beings are remarkably adaptable and can tolerate a wide variety of environmental conditions. In fact we are far more resilient than many products and processes, which have absolute environmental thresholds past which they are significantly affected, the first air-conditioning system, designed by Willis Carrier in 1902, was devised not to provide comfort for people, but to stop paper wrinkling under humid conditions in a printing plant.

So, given that comfort is a personal perception and that people can adapt to a wide range of conditions, how do we define the threshold at which an environment becomes unacceptably hot?

The study of overheating is a relatively new and developing field of science. CIBSE established a specific overheating task force in 2007, which published a guidance document (KS16 How to manage overheating in buildings) in 2010, just as the first tranche of Design for Future Climate projects got under way. While this was primarily focused on non-air-conditioned office buildings, it provides an excellent overview of the factors that contribute to discomfort, much of which can be applied to air-conditioned offices and to other building types.

The task force has continued to support the development of a new Technical Memorandum (TM52 The limits of thermal comfort: avoiding overheating in European buildings), which incorporates the latest research. This is scheduled for publication in early 2013, and the thinking behind it will be incorporated in the eighth edition of CIBSE Guide A, to follow later in the year.

Some of the project teams referred to the adaptive comfort approach (discussed in detail below) on which this new guidance is based, but the methodology was only applied on a very small number of projects. The majority used overheating and comfort metrics taken from the seventh (2006) edition of CIBSE Guide A or other recognised standards current at the time. Their studies do, however, provide useful insight into the levels of adaptation that will be necessary and the kind of solutions that are available.

CIBSE Guide A (2006) provides recommended summer and winter operative temperatures for buildings with full year-round temperature control. These stem from the current thermal-comfort standard BS EN ISO 7730, which is based on laboratory studies of how individuals react to different temperatures. The methodology was developed by Ole Fanger at Kansas State University in the 1960s, based on a "Predicted Mean Vote" (PMV) or "Percentage People Dissatisfied" (PPD), reflecting the likely votes of a large group of people on whether a space is comfortable, too warm or too cool. This was used to determine a range of conditions that most building occupants will generally find comfortable, with the PMV or PPD calculated empirically from the personal and environmental variables above.

Buildings with mechanical cooling and heating systems can be engineered to provide thermal comfort within the narrow range defined. However, Guide A also recognises that this approach may not be always applicable to free-running buildings, in which higher internal temperatures may be generally acceptable in summer.

Operating a naturally ventilated building is rather like sailing a yacht. Regular adjustments must be made to keep it on course – lowering a shade to reduce solar gain, opening a window for greater ventilation – but some variation is expected and, indeed, may enhance the perception of comfort. A mechanically ventilated building is more like a passenger liner; the course is corrected by the crew and there is greater expectation of a smooth passage.

Guide A reflects this difference to an extent by offering values for generally acceptable indoor summer temperatures for naturally ventilated buildings, above which some people will begin to feel warm, as well as a threshold temperature. If this benchmark is exceeded for a stated percentage of occupied hours, the building is considered to overheat (see table below).

BUILDING TYPE		MECHANICALLY VENTILATED	NATURALLY VENTILATED	
		Operative summer temperature	Operative temperature for indoor comfort in summer	Benchmark summer peak temperature (should not be exceeded for more than 1% of annual occupied hours)
Offices		22–24°C	25°C	28°C
Schools		21–23°C	25°C	28°C
Dwellings	Living rooms	23–25°C	25°C	28°C
	Bedrooms	23–25°C	23°C	26°C

● ● ● **3.3** Selected CIBSE summer comfort criteria for mechanically and naturally ventilated buildings.
Source: CIBSE Guide A (2006), tables 1.5, 1.7 and 1.8

Multiple criteria

Summertime comfort criteria for new schools are set out in Building Bulletin 101 (BB101). This requires buildings to be analysed for occupied hours (09.00 to 15.30, Monday to Friday) for the period May 1 to September 30, using the geographically closest current CIBSE Test Reference Year, and sets three criteria, two of which must be met:

- maximum of 120 hours when classroom air temperature rises above 28°C
- average internal-to-external temperature difference should not exceed 5°C (ie the internal air temperature should be no more than 5°C above the external air temperature on average)
- internal air temperature during occupation should not exceed 32°C.

There are, however, a number of problems with the above criteria.

- Setting a static threshold, even through the summer months, does not take into account the relationship between acceptable internal temperatures and changes in external temperatures. In a warmer future climate, people are likely to become accustomed to higher temperatures.

- The first criterion measures only the occurrence of overheating, rather than its severity. By this measure, a building where the temperature exceeded the threshold by a small amount for a long period would be deemed to have a more serious overheating problem than one where the temperature shot well over the threshold but for a relatively short time.

- An annual average can conceal extremes during certain months and does not reflect concentrated periods of higher temperatures, which are arguably more significant in perceptions of thermal discomfort. Averages over a whole building also hide problem areas – on elevations exposed to the sun (south- or west-facing), for example.

- A sharp threshold is more vulnerable to assessment methods and calculation errors; there is also the potential to "solve" an overheating problem by simply extending the number of occupied hours.

Adaptive Comfort Threshold methodology

For naturally ventilated buildings without cooling, an "adaptive" approach to summer comfort can be taken. Field studies have demonstrated that people are more tolerant of temperature changes than laboratory studies suggest. This is because they adapt to changing conditions over time, and consciously and unconsciously act to regulate their thermal environment – by altering their activity level or clothing, for example, or by opening a window or closing a blind. Studies show that the temperature at which most people feel comfortable inside changes with the average outdoor temperature, and that higher internal temperatures will be acceptable in summer than in winter.

The adaptive approach is embodied in the Adaptive Comfort Threshold (ACT) methodology (fully described in BS EN 15251) which determines a varying running mean internal comfort temperature which tracks the external temperature, taking account of conditions over previous days.

Arup used the ACT methodology on two projects: Church View, Doncaster, and the University of Sheffield Engineering Graduate School.

The graph below, produced for Church View, shows how the ACT values change through the year and through the century (based on CIBSE TM48 Medium-high emissions weather files) as compared with the traditional 28°C peak temperature target. It also illustrates that internal temperatures of 28°C in August might be acceptable, while the same temperature in June would not.

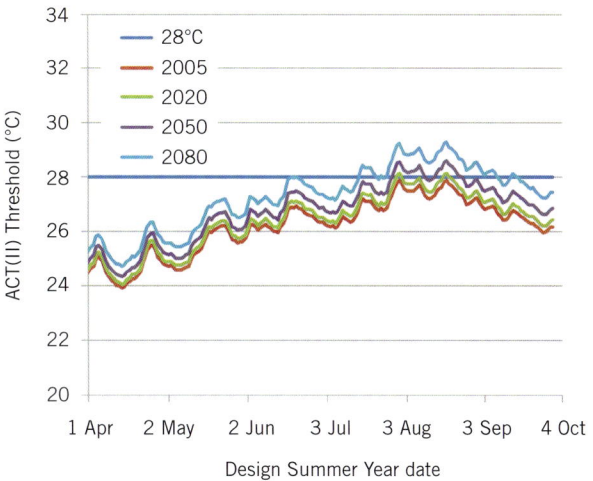

3.4 This graph, produced by Arup for the Church View project, shows how the Adaptive Comfort Threshold could vary over the century.

For the University of Sheffield project, Arup analysed the PROMETHEUS weather files (50th-percentile, Medium emissions) to show that the maximum ACT is currently 28°C, but that by the 2080s this temperature would be considered comfortable for 76 days.

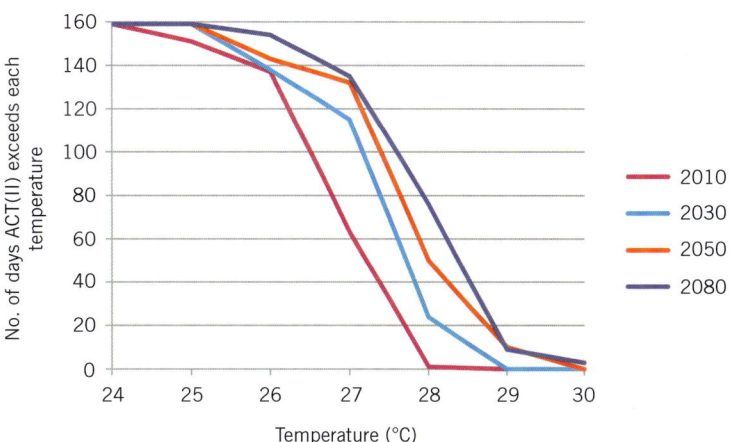

3.5 Arup produced this graph for the University of Sheffield project to show that temperatures above 28°C would be considered comfortable for an increasing proportion of the summer in Sheffield later in the century.

They proposed three overheating criteria, two of which would need to be met for the building to be considered not to overheat:

1. Hours of Exceedance: the internal temperature should not exceed the ACT by more than 1°C for more than 40 occupied hours per year
2. Weighted Exceedance: this should be a maximum of 10 days, based on a combination of the duration of overheating and its severity
3. Upper temperature limit: the peak internal temperature should not exceed the ACT by more than 3°C.

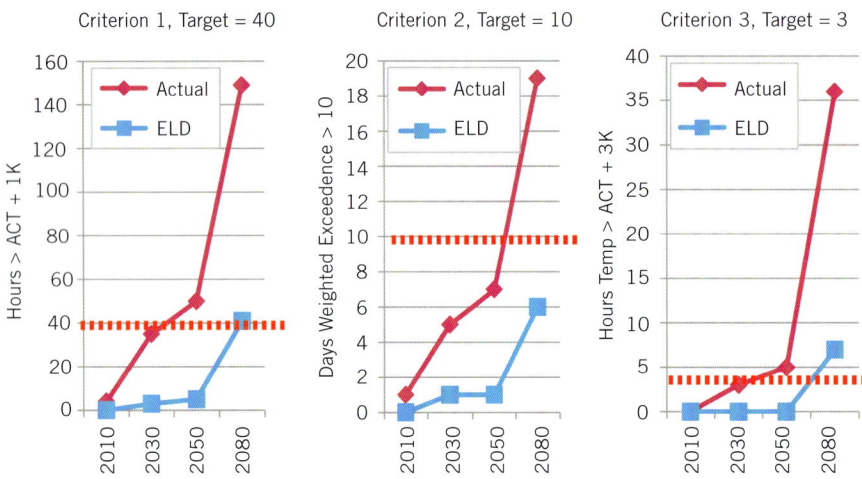

3.6 A comparison of the three metrics within the ACT methodology proposed by Arup for the University of Sheffield project, comparing the performance of the actual building and the team's Engineering Led Design (ELD).

Arup chose to use the first criterion as the main metric for both projects. But for the Sheffield project, they also compared this with the others, for both the actual building and the alternative Engineering Led Design (ELD) produced for the Design for Future Climate project (see figure 3.6).

The new CIBSE TM52 proposes the ACT methodology as a more robust approach to defining overheating than the PMV model, particularly for mixed-mode buildings or for buildings without mechanical cooling. In addition, recent findings indicate that, even in a fully mechanically ventilated building, there is a correlation between acceptable internal comfort and external environmental conditions.[7]

The methodology is clearly somewhat more complicated to use than the simple absolute criteria used by most of the project teams, but it does acknowledge the limitations of a single indoor temperature limit irrespective of outdoor conditions. It also inherently includes the capability to allow for climate change – taking some account of our ability to adapt to generally higher temperatures, as well as seasonal variation and hot spells.

APPROACHES TAKEN BY THE PROJECTS TEAMS

A number of the teams designing sealed, mechanically ventilated buildings took the position that they would generally continue to provide comfortable conditions, subject to the capacity of the cooling system, which could be sized or upgraded accordingly. In some cases they used overall energy consumption as their principal metric to judge how effectively proposed strategies would counteract rising external temperatures. The question they set themselves was not "will the building remain comfortable in future?" but "will it become unsustainably energy-hungry?"

No doubt the teams did analyse comfort conditions in different parts of the building when they designed the cooling system for their live project. However, it is worth stressing that aggregated metrics such as total energy consumption reveal none of the finer details of problem areas where overheating may be occurring. Even for a straightforward open-plan, repetitive building, it is important to compare the perimeter with the core, for example, and to consider how different sides of the building will be affected.

The heat maps produced by AECOM for the EBI (European Bioinformatics Institute) Hub and the London School of Hygiene and Tropical Medicine are a neat way of analysing the outputs from a comprehensive model of individual spaces in a building. This not only picks up specific areas that are overheating, but also indicates the magnitude of the problem in each location.

3.7 The Technical Hub or EBI2 building at the Genome Campus, Hinxton, Cambridgeshire.

● ● ●   USING HEAT MAPS

These heat maps were prepared by AECOM for the London School of Hygiene and Tropical Medicine to show the results of the modelling for all of the rooms in the building.

Each room is represented by a row in the table, arranged by orientation. Across the tops of the columns are the temperature thresholds used in the modelling. Cells are coloured red where the given temperature is exceeded for more than 1% of occupied hours, and blue when it is never exceeded. Shades of orange and yellow indicate various points between 0 and 1%. As there are both naturally and mechanically ventilated spaces within the building, two approaches were taken: to test for overheating, and for set-point exceedance.

OVERHEATING MAPS

The first three tables show the incidence of overheating for 48 office and administrative spaces which are heated only. For this modelling, the dry resultant temperature of the air in the spaces was monitored, and a temperature range of 26°C to 34°C was used. There is a red line running vertically at the 28°C point to indicate when the rooms are hotter than the CIBSE benchmark.

The table on the left shows that overheating is a significant problem in the current climate. As the century progresses, all but a few of the spaces will overheat to an unacceptable level for more than 1% of occupied hours.

(a) Baseline   (b) 2050s   (c) 2080s

3.8 These heat maps, produced by AECOM for the London School of Hygiene and Tropical Medicine, show the increasing number of hours when rooms in different parts of the building are projected to overheat, under current conditions and in the 2050s and 2080s.

SET-POINT EXCEEDANCE MAPS

For 30 mechanically ventilated spaces, AECOM modelled the percentage of occupied hours when the air temperature rises above the cooling set point, using a temperature range of 22°C to 30°C. The set point is 24°C, meaning that the cooling system starts when the air temperature exceeds this limit, so the room temperature would be expected to fluctuate at around 24–25°C. The team noted that though an air temperature above the set point does not necessarily represent immediate discomfort, it can be progressively more uncomfortable in spaces where expectations of comfort are higher. The maps show that cooling systems sized for today's climate will increasingly struggle in the future, with set-point exceedance worsening steadily through the century.

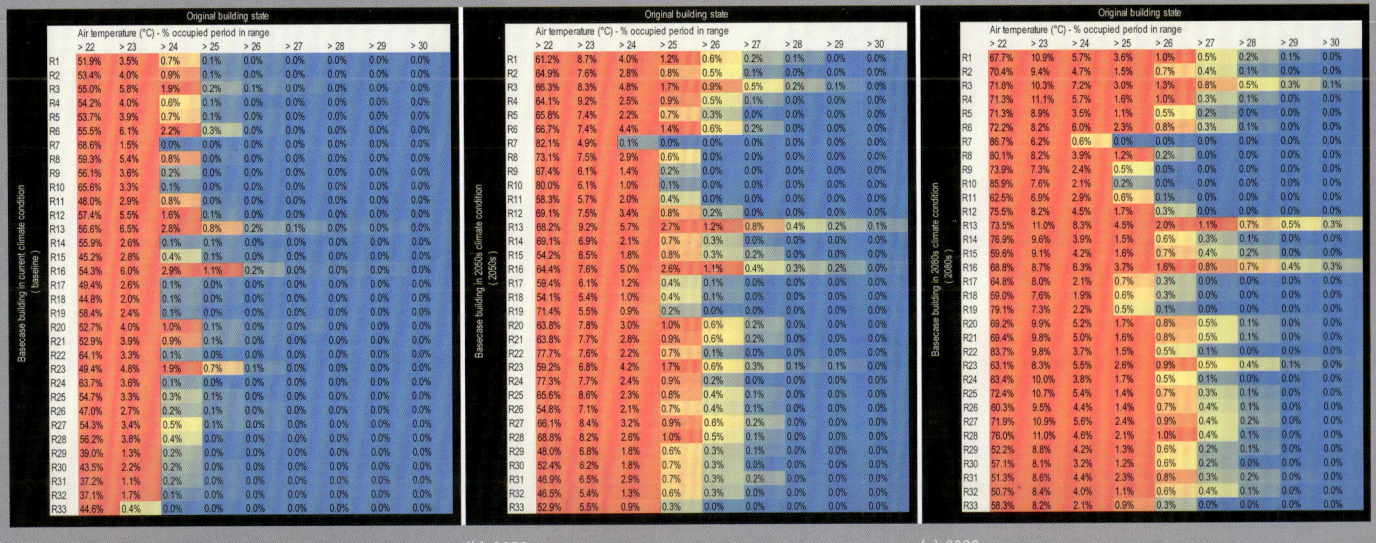

(a) Baseline   (b) 2050s   (c) 2080s

3.9 These heat maps, produced by AECOM for the London School of Hygiene and Tropical Medicine, show the increasing number of hours when cooling set points are exceeded for rooms in different parts of the building, under current conditions and in the 2050s and 2080s.

COOLED-SPACE OVERHEATING MAPS

AECOM also prepared specific overheating maps for the mechanically cooled spaces, similar to those for the naturally ventilated spaces, using the dry resultant temperature.

The maps show that while spaces do not overheat to the same degree as the naturally ventilated spaces, some will become hotter than the 28°C benchmark for significant amounts of time by the 2080s.

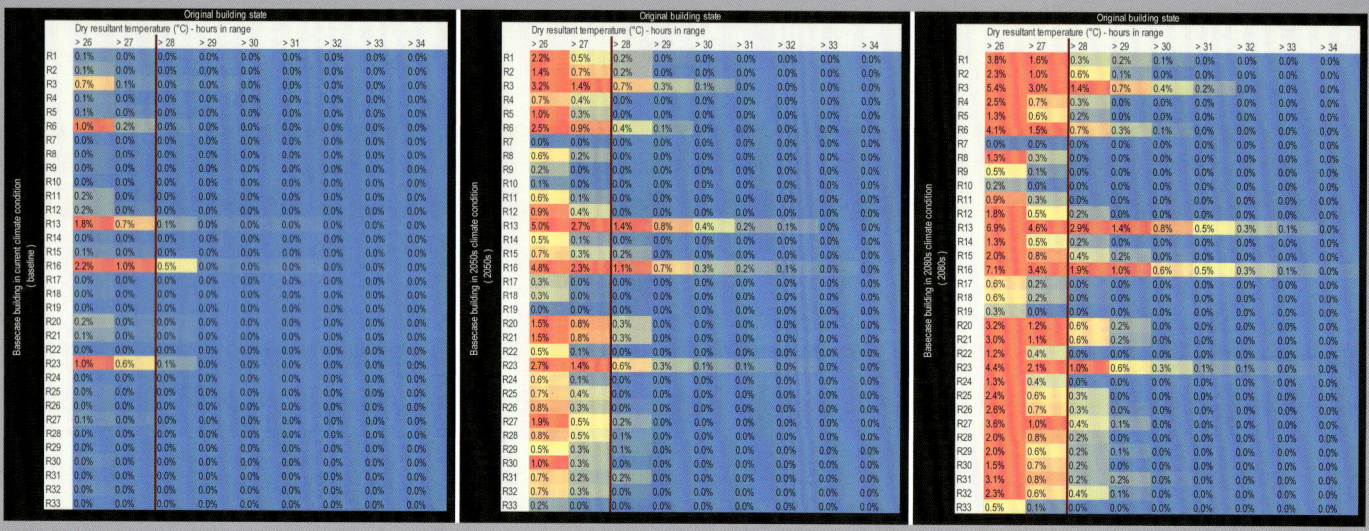

(a) Baseline   (b) 2050s   (c) 2080s

3.10 Overheating maps, produced by AECOM for the London School of Hygiene and Tropical Medicine, showing potential overheating of mechanically cooled spaces.

3.11 Keppel Street building, London School of Hygiene and Tropical Medicine.

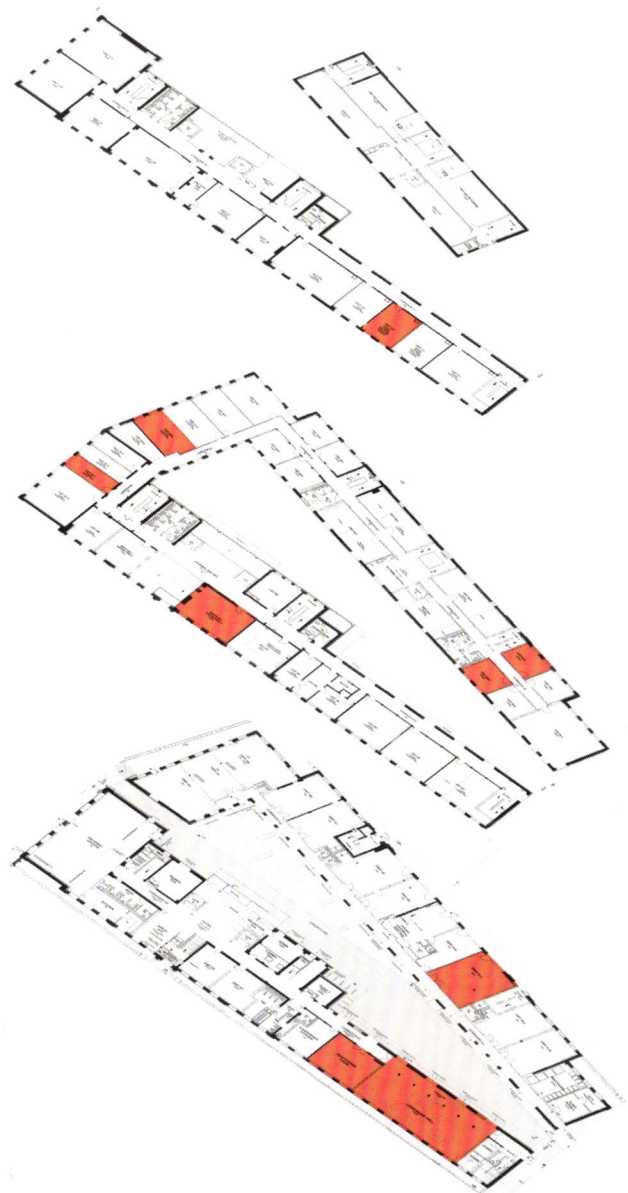

3.12 For the Church View office conversion project in Doncaster (above), designed by Bauman Lyons, the team chose a set of sample rooms to assess the impact of future climate, highlighted in red on a plan of the building (left).

Rather than carry out an all-encompassing heat-map-type review of every space in their building, the team working on Church View, an office conversion in Doncaster, used their combined experience to select a set of spaces with varying orientations and characteristics, which they used as a sample to evaluate the effect of projected changes and to test overheating solutions in some detail.

This seems a particularly appropriate technique for the team's incremental room-by-room approach to combating overheating in an existing building. For each space, they produced a tailored strategy that could be implemented according to the characteristics of the spaces and how they are used.

Identifying the issues associated with particular areas – whether stemming from the use of the space or its thermal characteristics – allows designers to evaluate whether there is a general problem, such as with a facade or construction strategy, or whether the overheating is related to a particular use or location within the building. They can then consider general or specific ways of addressing the problem. For example, they could group certain uses or move them to less stressed locations in the building.

● ● ●
COMFORT STANDARDS FOR SCHOOLS

Engineer Max Fordham raised a number of issues with the school-building standard Building Bulletin 101 when working on the Worcestershire primary schools project.

The project encompassed three schools, including an existing 1960s lightweight building, which was used as a basis for adaptation work. The team decided to involve the users in establishing suitable overheating criteria. They found the children were extremely perceptive in identifying the occurrence and causes of overheating in rooms that might have been classed as performing adequately against the standard – raising the question of whether the standard itself is satisfactory.

Their analysis highlighted the following limitations of BB101 (described on page 45).

- The threshold temperatures are the same UK-wide, allowing a greater difference between internal and external temperatures in cooler regions.
- It uses the Test Reference Year rather than the (usually) warmer Design Summer Year.
- It applies only to the building's occupied period, which does not take extended uses such as after-school clubs into account. There is a greater risk of overheating in the late afternoon because rooms have been warming up throughout the day.
- Cumulative measures of overheating mean that a room could be 27°C consistently and technically pass, but still be unacceptably hot.
- The criterion that the *average* internal temperatures should not exceed the external by more than 5°C allows cool temperatures towards the beginning of the day to offset peaks towards the end, which again means that a classroom that is uncomfortably warm in the afternoon could still pass.

Following this analysis, it was decided to aim for the following, tougher, summer comfort criteria:

- to use a DSY rather than a TRY
- maximum of 100 hours above 25°C
- maximum of 20 hours above 28°C
- zero hours above 30°C.

Of course, all these overheating criteria depend on a chain of factors falling into place: the accuracy of the inputs to the designers' dynamic simulation models, the accuracy of the algorithms in the models themselves, the survival of essential components through the inevitable "value engineering" process, good site workmanship and controls that can be used by the eventual users of the building – and all this before taking into account the uncertainties inherent in climate projections.

In order to build confidence in our adaptation responses, it is essential that detailed on-site monitoring of how buildings behave in reality (such as the work being carried out under the TSB's Building Performance Evaluation programme) is cross-checked against the models that were used to design them.

4 DESIGNING FOR THERMAL COMFORT: Controlling Heat Gains

Low-energy building design is a balancing act. On the one hand, we must minimise heat loss and exploit any incidental heat gains in winter, while excluding unnecessary gains in summer and allowing unwanted gains to escape. The aim is that the building does the majority of the work of providing comfort, leaving as little as possible to be done by supplementary mechanical heating or cooling systems.

In our relatively cool northern-European climate, our focus has been on minimising energy consumption during the heating season. It is now possible to design new buildings that are so well insulated and airtight that we can use solar and internal incidental gains to provide most of the heat we need to keep them warm. This restricts the period when top-up heating is needed to the depths of winter. But the corollary is that outside this short period, heat from these incidental gains needs to be allowed to escape in order to maintain a comfortable balance. This is relatively easy in mid-season when external temperatures are low, but in summer it may become a challenge to lose enough heat to avoid overheating.

In a warming climate, the balance is changing. We need to expand our understanding of the causes of summer overheating and develop robust strategies to minimise the need for supplementary cooling. Clearly, the most efficient way to reduce overheating is to reduce the gains that cause the problem in the first place. The balance between the various inputs will differ according to the building's location, form, orientation, construction and use. This balance will also vary between different spaces in the same building, and is not static – it will change throughout the day or week.

All the TSB project teams used dynamic simulation tools such as IES and TAS that take account of these factors. However, the inherent complexity of these sophisticated "black boxes" (whose

●●● BALANCING HEAT GAINS AND LOSSES

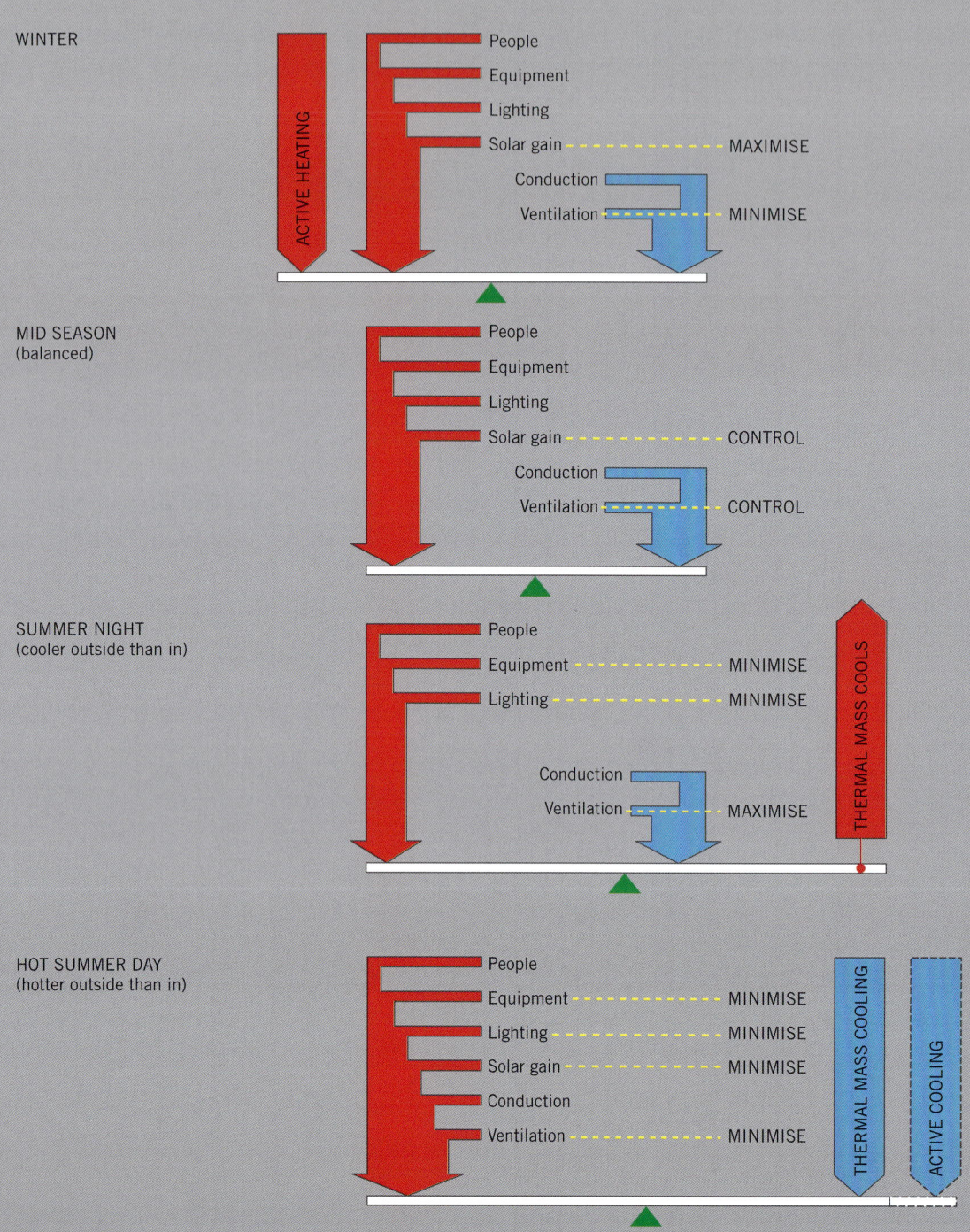

WINTER

ACTIVE HEATING

People
Equipment
Lighting
Solar gain — — — — — — — MAXIMISE
Conduction
Ventilation — — — — — — — MINIMISE

MID SEASON
(balanced)

People
Equipment
Lighting
Solar gain — — — — — — — CONTROL
Conduction
Ventilation — — — — — — — CONTROL

SUMMER NIGHT
(cooler outside than in)

People
Equipment — — — — — — — MINIMISE
Lighting — — — — — — — MINIMISE
Conduction
Ventilation — — — — — — — MAXIMISE

THERMAL MASS COOLS

HOT SUMMER DAY
(hotter outside than in)

People
Equipment — — — — — — — MINIMISE
Lighting — — — — — — — MINIMISE
Solar gain — — — — — — — MINIMISE
Conduction
Ventilation — — — — — — — MINIMISE

THERMAL MASS COOLING
ACTIVE COOLING

4.1 These diagrams show the key inputs and outputs of heat to
and from a building, and how they must be balanced at different times.

internal calculations are inaccessible to the user), and the fact that they are usually controlled by somewhat remote specialists, can obscure some relatively simple observations about the timing and relative magnitude of potential gains, which could direct architects' fundamental design approach. One very simple technique would be to plot a timeline over 24 hours for a given space, showing the heat inputs as they occur (people, primary use of equipment, standby equipment gains, lighting, thermal storage, solar gain, etc). This should enable us to see, very simply, the orders of magnitude of each source of heat, and thus what to focus on and where solutions might lie.

The Design for Future Climate projects explored a great range of strategies for minimising overheating. These are described below in sections organised around the sources of overheating, because the sources and their solutions are so closely linked.

●●● INTERNAL GAINS

Heat gains from the occupants of a space vary according to the density of occupation and the activities they are engaged in.

Any device that uses energy in a space will also produce heat. Even in a domestic environment, this can make a significant impact, as architect Gale & Snowden realised on the Extra Care 4 Exeter project. In considering ways to reduce internal gains, the team recommended that there should be no hot-water storage in flats, and that heat-generating activities such as cooking,

4.2 Extra Care 4 Exeter scheme, designed by Gale & Snowden.

tumble-drying and ironing should be kept to a minimum, particularly when temperatures were extreme. Occupants who required a cooked meal would be encouraged to use the cafe in a separate building.

Teams working on office projects noted changing patterns of occupation and technology use that are likely to affect internal gains in future. These appear to "pull" in different directions – with denser occupation and more gadgets increasing internal loads, countered by improvements in the efficiency of lighting and IT equipment.

To reduce the heat that artificial lighting produces, daylighting should be exploited where possible. However, simply increasing the glazed area does not necessarily increase the amount of useful daylight, and it can just result in increased solar gain. To be useful, the light must reach the right places without causing glare, requiring considered design – particularly in how it is controlled as the sun moves around the building through the day and over the seasons.

On the Church View project in Doncaster, a managed workspace for the creative and media industries, architect Bauman Lyons and engineer Arup considered the trend for much denser occupation and longer working hours, which will have an increasingly significant effect on internal heat gains by the 2080s. This shift is especially pronounced in more expensive city-centre locations, and particularly so for the small and micro businesses (<50 employees) that the project is intended to house, where intensive, 24-hour working as an inherent element of global business is starting to challenge conventional patterns of use.

The Church View team also examined the potential of adjusted lighting levels and emerging technologies, such as LEDs, in their assessment of the impact of internal gains. The graph below shows the effect of reducing the lighting gains from 12W/m² (current good practice) to 6W/m², representing a reduction in lighting levels to 200 lux of background lighting plus task lighting.

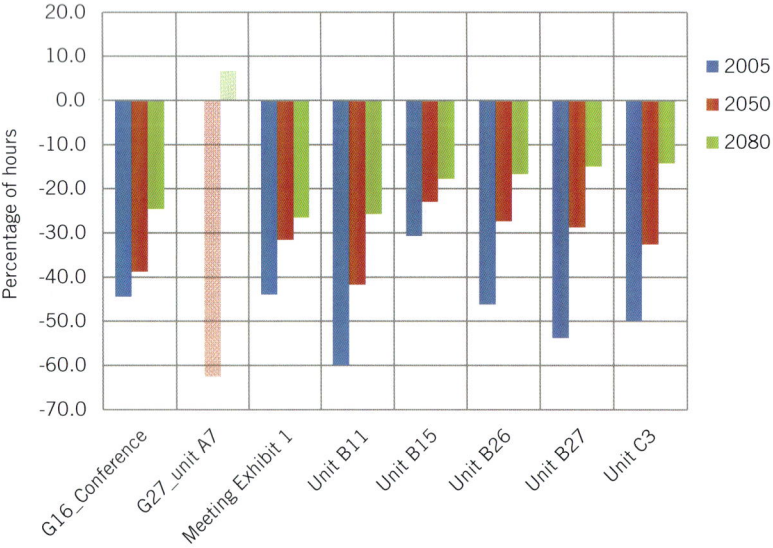

4.3 The graph was produced for the Church View project to show the percentage reduction in overheating hours that could be achieved by reducing heat gains from artificial lighting, for eight different rooms through the century. Although there is an effect for all years modelled, it is most pronounced for the 2005 data.

They also reviewed future ICT trends, which are likely to lead to sweeping changes in working practices, networks, and hardware and software. They suggested that over the next quarter-century the concept of working in an office, with a desk for each user, will become a thing of the past, and that businesses will increasingly cease to own their ICT equipment and software in favour of buying-in ICT services and capacity, drastically reducing the space, power and cooling requirements associated with ICT within buildings. At the user level, there will be less need for local processing power, and desktop PCs will be replaced by "thin-client" systems, simpler devices that produce less heat.

Their modelling showed that reducing heat gains from IT equipment significantly reduced the number of hours when the building would overheat. The graph below shows the impact of upgrading to more efficient IT equipment, from a current base case of 15W/m² to 5W/m² through moving from PCs to laptops or thin-client computing, and more efficient printers and peripherals.

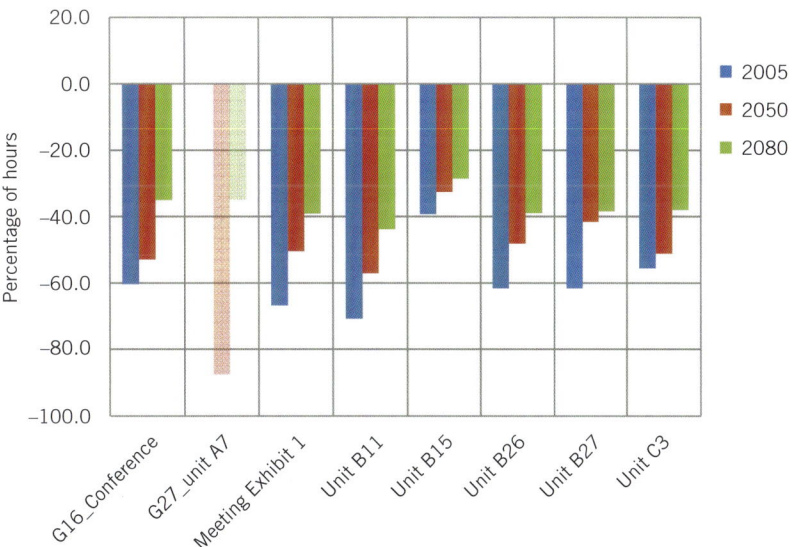

4.4 This graph shows the percentage reduction in overheating hours that could be achieved by reducing heat gains from IT equipment, for the eight sample rooms of the Church View project. As with reducing artificial lighting loads, it is generally effective for all spaces.

However, claims of energy (and, therefore, heat-production) savings from thin-client systems need to be interrogated thoroughly. It may be that replacing larger loads from PCs that run for a limited number of hours (provided they are switched off when not in use) with smaller but continuous loads adds up to a similar total in the end.

Arup was also the engineer on the 100 City Road project, where the team suggested that daylight-controlled dimming could reduce lighting energy by as much as 30%, and that similarly providing 200 lux supplemented with task lighting, as opposed to the current BCO (British Council for Offices) standard design of 400 lux, could save 15% of the building's energy demands.

They also estimated that reducing the IT equipment load using thin-client technology could cut the energy demand for cooling by 10% (as well as reducing the electricity used by IT systems), though they noted that predicting energy savings from these systems is dependent on usage patterns. More heating would be required in winter as a consequence, but they anticipated this would be reduced in future with warmer winter temperatures. In the short term, using gas to provide the shortfall in heat is inherently more carbon efficient anyway. The building was also planned with distributed server rooms, so that useful heat could easily be recovered from them during the heating season.

The new Admiral Insurance headquarters in Cardiff has a moderate-to-deep floor plan with a central core, typical of many office buildings around the country. Modelling showed that the effect of rising external temperatures was relatively insignificant compared with internal gains, and that changing occupancy levels would have a much more significant impact on cooling demand than fabric adaptations such as increased shading and thermal mass. A reduction in occupancy levels (and the associated use of equipment) from 5.9m²/person to 6.5 m²/person would result in a significant decrease in demand for cooling, with little effect on heating demand.

4.5 Admiral Insurance headquarters, Cardiff.

Even by 2080, when demand for cooling was shown to have increased somewhat, lighting and equipment accounted for by far the most significant proportion of energy consumption. This led the Admiral team to recommend that all lighting be changed every 10 years as a win-win strategy to save lighting energy and reduce the cooling needed to remove the heat from less efficient fittings.

The project engineer, Hoare Lea, also predicted that the equipment gains from office apparatus would drop significantly: from 25W/m^2 in 2015 to 10W/m^2 in 2080 and that lighting energy consumption would similarly drop from 9W/m^2 in 2015 to 3W/m^2 by 2075.

In the shorter term, they suggested that the client consider converting to an IT system that runs on direct, rather than alternating, current. A network of DC-powered computers installed at the University of Bath library was found to emit half as much heat energy as the previous AC-powered system.

Changing behaviour – soft adaptation

We must also bear in mind that the climate is not the only thing that is changing and that there may be "soft" adaptation measures that are very effective alternatives to investment in building fabric or services. For example, ubiquitous mobile connectivity gives us the option to work at home, in the park or, if the opportunity exists, just to step outside and work under a tree when the office gets too hot.

The team working on the University of Greenwich project contemplated the potential benefit of a change to working hours. They observed that with the mixed-mode system currently installed, mechanical ventilation is typically switched on from 11am until 5pm. If the university were to introduce a siesta-type regime whereby the facilities are closed from 3pm until 5pm and the workday extended into the evening, energy use could be significantly reduced. During July in the 2080s, savings were estimated to be in the region of 10,000kWh. Without the siesta, CIBSE comfort levels were exceeded by 2020, whereas this could be delayed until sometime between 2020 and 2040 if it were introduced.

The team working on the rationalisation of council offices in Cornwall considered the impact of more flexible working hours, increasing the length of the possible working day and giving staff the option to work early or late, and thus reducing occupancy levels at times of day when external environmental conditions were at their most severe. However, this was found actually to increase both energy usage and overheating, because the prolonged heat gains from people and lighting would be greater than the reduction achieved by reducing peak occupancy.

Shading

The vulnerability of unprotected glazing to solar gain depends on its orientation. North-facing windows are subjected to relatively little solar gain compared with other orientations; however, it is worth noting that in midsummer it is through east- and west-facing windows that maximum peak solar radiation will enter (at a level almost 50% higher than through those facing south). In mid-season, the peak radiation levels are roughly the same for south-east, south and south-west windows, when 2m² of window can produce as much heat as a one-bar (1kW) electric fire. The graphs below are taken from CIBSE Guide A (2006) and show how solar cooling loads change throughout the day and through the seasons.

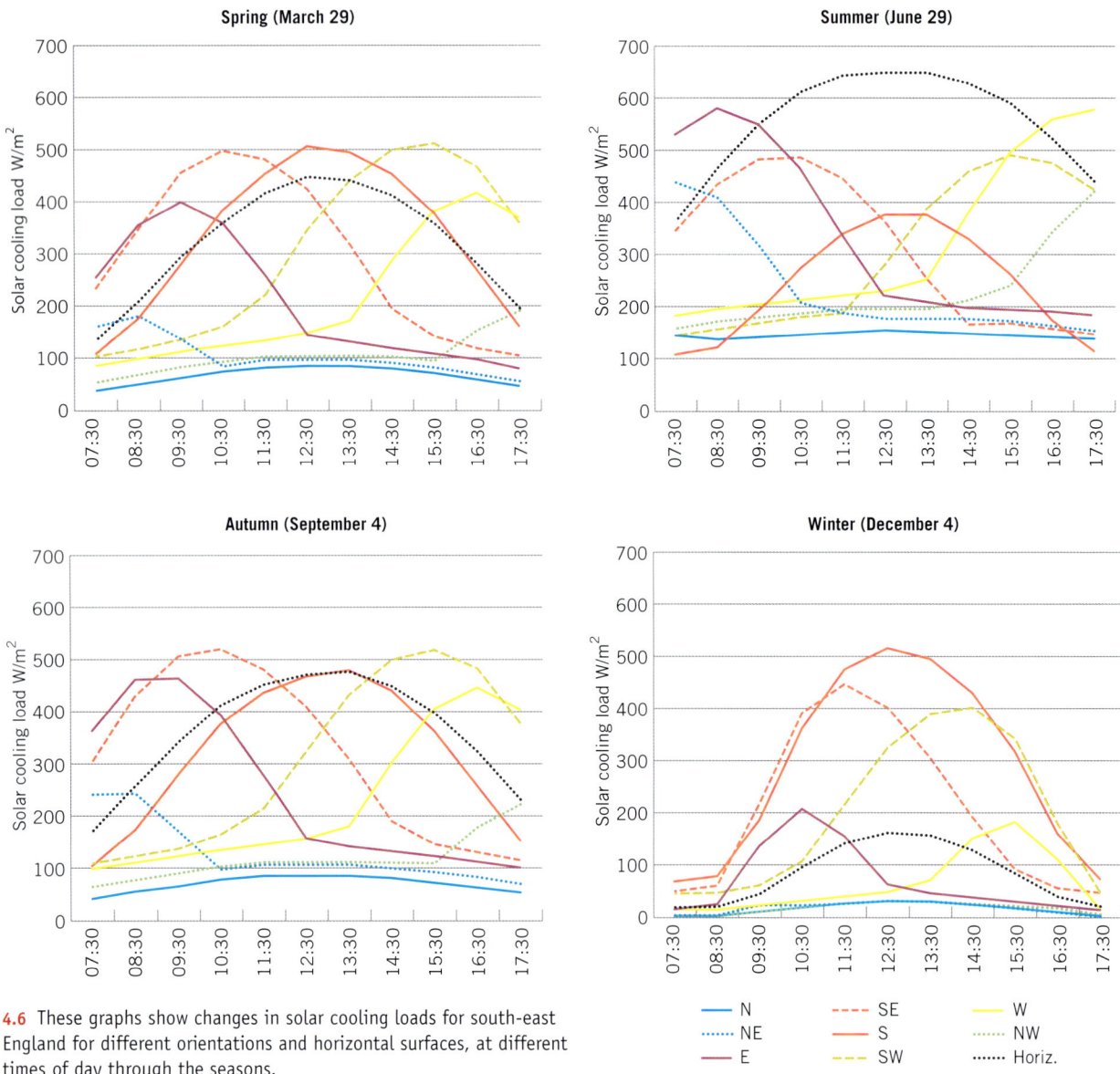

4.6 These graphs show changes in solar cooling loads for south-east England for different orientations and horizontal surfaces, at different times of day through the seasons.

Source: CIBSE Guide A (2006), table 5.19

Even experienced teams can be caught out by unanticipated solar gains. The EBI Hub team spotted that some east-facing rooms in the existing design were unexpectedly overheating early in the morning before the cooling system had come into operation, even though the facade had a fairly elaborate shading system. This shading was adjusted to deal with the problem.

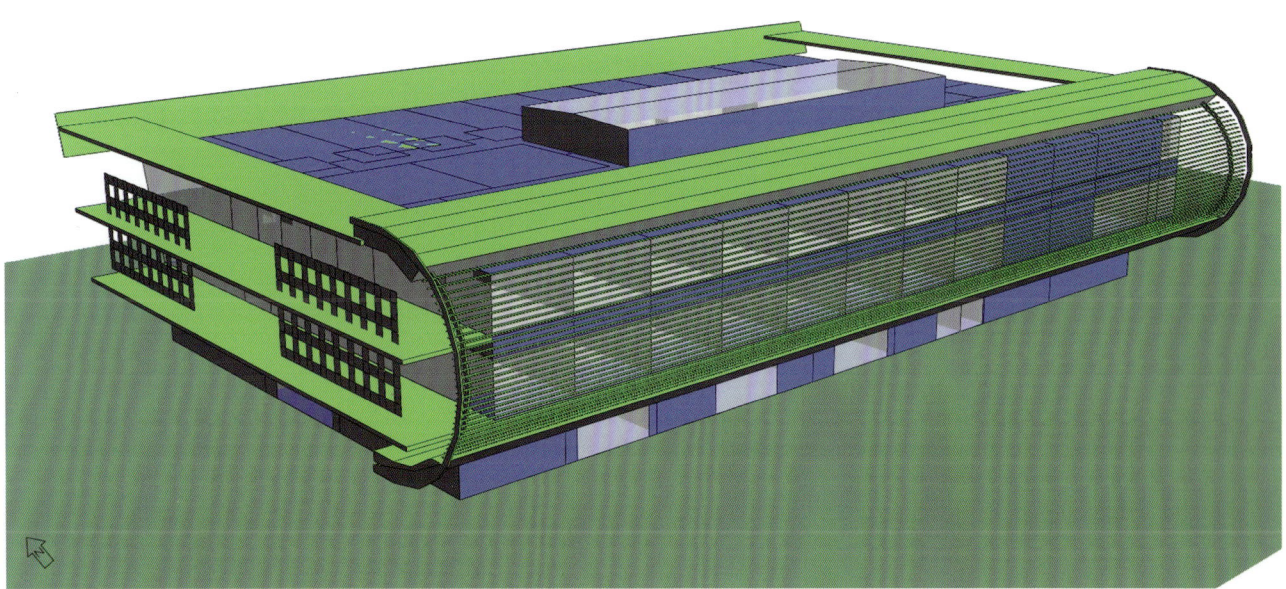

4.7 An IES model of the EBI Hub, showing louvres on the south facade and vertical blind mesh on the western facade.

On south-facing glazing, sun angles are high in the summer and low in the winter so it is relatively easy to shade against excessive solar gain in summer, while still providing good daylighting and some useful winter gains. It is more difficult to achieve a satisfactory balance with east- and west-facing glazing because the sun hits at a lower angle all year round: the aims of letting in daylight and excluding solar radiation are therefore incompatible. "Blinker" shading – consisting of vertical shades projecting at right angles to the glazing, which shade it from oblique, southerly sun angles for as long as possible throughout the day – can go some way to reducing the problem. But at some point in the day an opaque barrier will be necessary, even with solar-control glass, if only to control glare.

Movable shutters are a potential solution. These could be used when needed in summer and at particular times of day but folded away in winter, and when the facade is not in full sun, to maximise usable daylight. They are the norm in more southerly climates, where they are highly effective because their use is well understood. In the UK, we have no such tradition and would need to learn how to operate them properly.

● ● ●   EXTERNAL SHUTTERS

For the Church View refurbishment project, Bauman Lyons considered the use of standard European-style horizontal, slatted roller shutters, which are flexible and effective in operation and discreet when retracted. However, it was not possible to incorporate them into the existing construction. As an alternative, they developed a bespoke, lockable folding-shutter system, comprising two pairs of side-hung bi-folding shutters to the upper part of the window and one pair of horizontal bi-folding shutters to the lower part. This arrangement would allow the larger, upper part to be closed whilst still allowing the lower part of the window to remain open. The horizontal louvres in the shutters were all adjustable for fine tuning. Unfortunately, fears that local planners would object meant that the ideas were not taken further.

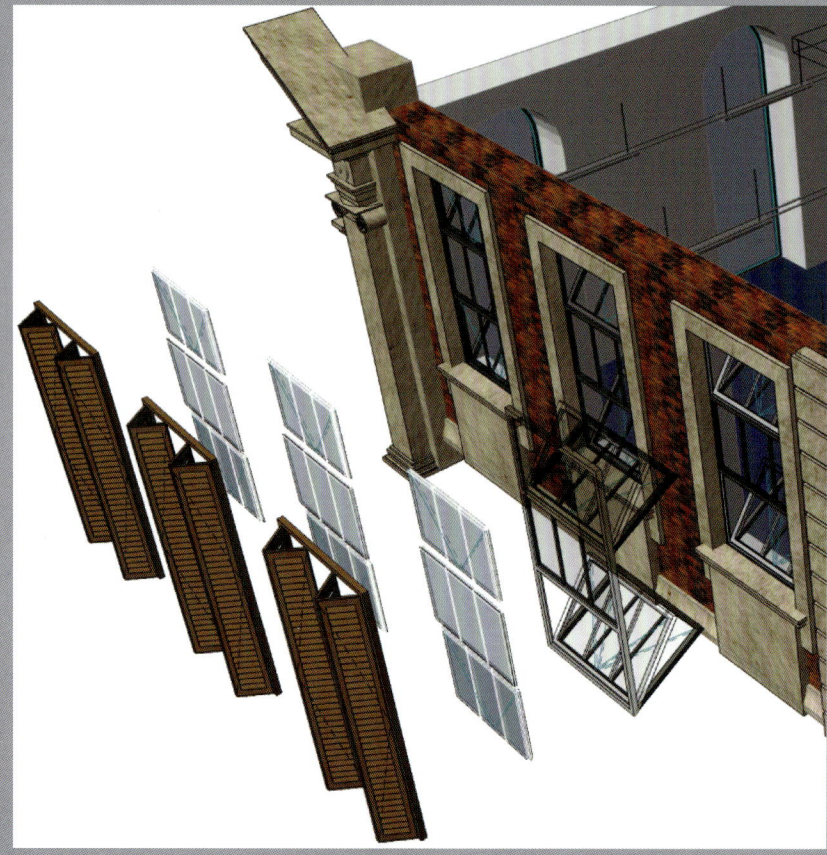

4.8 Bauman Lyons designed a bespoke shutter system for the Church View project.

Shading can provide other benefits: the Worcestershire schools project proposed external pergolas that both shaded south-facing windows and defined external shaded teaching spaces outside classrooms. This approach is much more affordable than the alternative option of replacing existing glazing with solar-control glass, and could be constructed as a parents' DIY project executed in stages as funds were raised.

4.9 External pergolas proposed for Franche Primary School, Worcestershire, could provide not only shading, but also external teaching spaces, as in this visualisation by Sjölander da Cruz Architects.

For the purposes of analysis, the effect of planting the pergolas was ignored. However, deciduous planting in applications like these offers multiple benefits – being cheap to implement, delightful and tuned to the seasons to allow maximum daylight and solar gain in winter while controlling summer overheating. Of course, even in winter, mature deciduous planting will affect daylight levels to some extent, and this needs to be taken into account.

Top louvre options
The top louvres provide direct shading to the glassed area of the south facade, they are not solid so as to allow air flow between them.

Timber slats fixed between an aluminium section at either end.

Polycarbonate screen to be used in all options diffuses light and provides shelter from rain.

Rounded aluminium aerofoil louvres fixed between aluminium frame.
Work/Play Seating module

Side shading options
To provide shading for external work play areas and circulation space within the canopy. But also a surface for signage or use of colour to give individual areas identity.

Tensile fabric screen fixed between post and primary structure. Possibly used for signage or for varying colours.

Vertical timber sections fixed at top and base and supported by intermediate steel rods.

Aluminium shading mesh within a timber/steel frame.

Description
The drawing shows the progression and the layering of materials and components to achieve the complete unit. We have provided various options for the different components which will allow us to find the optimum solution for the canopy in tems of shading, air flow, enclosure and useable space. The structure remains constant while the other elments become a 'kit of parts' that fit into place and are interchangable.

Work/Play Seating module
The module extends the curriculum space by making use of the space, shelter and shading provided by the canopy and planting extension. Seating, storage and work surfaces are incorporated into the form.

Planting Extension
The vertical and horizontal climbing plants again extend the external usable space; they provide habitat for insects and plantlife, act as a learning tool and clean the air.

4.10 Exploded diagram of the external pergolas designed by Sjölander da Cruz for the Worcestershire primary schools project.

A number of teams considered the potential for deciduous tree planting to provide seasonal shading very economically, with the shading effect increasing as the trees grow and the climate changes. However, it is not generally possible to analyse the shading effect of trees with any precision using current versions of dynamic simulation tools. This was a stumbling block for Bauman Lyons when assessing the potential for using trees to shade the heavily glazed south facade of the Church View project. The report noted: "It has not been possible to model the precise cooling benefits of these trees and, indeed, accurate modelling could only ever consider a single point in time which would be obsolete as trees grow, die and are replaced."

Any shading system will inevitably affect daylight levels. While a number of teams discussed the relationship between controlling solar gain and reducing daylight, none explored this in great detail. The Worcestershire schools team carried out some investigation into the effect of lowering the external pergolas to below the tops of the windows, allowing daylight into the classrooms at high level. This was found to improve daylighting levels, in comparison to completely shading the windows, without substantially increasing the tendency to overheat. High-level daylight penetrated deeper into the space while the areas adjacent to the window remained shaded, improving the uniformity of daylight generally. This analysis was carried out for a south facade – on other orientations, it is trickier to exclude low-angle sun while still providing good daylight.

Roof glazing via north lights is a very effective way of providing even daylighting while avoiding significant solar gain, though care should be taken that sunlight is not allowed to enter from other directions. The Worcestershire schools project included a study of rooflight options for a courtyard that had been roofed over. An arrangement of regularly spaced north lights provided good, even daylight without excessive solar gain; however, a more flamboyant central rooflight with glazing on all sides produced very uneven lighting and considerable solar gain – about 14kW, equivalent to 130 or so people in the space!

Solar-control glazing

A number of teams explored the benefits of using glazing with low solar transmission. As was noted by the Worcestershire schools team, this does not make sense in combination with solid shading. The two interventions are trying to do the same job, and will not represent good value for money if they are used together.

Generally, adding a window film or installing solar-control glass results in tinted windows. This represented an aesthetic opportunity for some project teams, but others felt that the tint created a feeling of dislocation between inside and outside. It becomes especially noticeable when the view through the tinted glass can be compared with that through an open window.

This was a particular issue for the Central Saint Martins project, an art college where both the quantity and quality of daylight is extremely important. Here, the team explored emerging nanotechnologies that hold the promise of selectively excluding specific ultraviolet (UV) wavelengths without affecting the colour of the light, using LCD (liquid crystal display) technology. There is even potential to tune the process so that the glass behaves differently in summer and winter, admitting more heat in the winter and excluding it in summer. Unfortunately the technology is still some way from commercial availability, though the team anticipated that it might have become established by the time the building requires its next major refit.

Internal blinds

WSP Built Ecology, working on Great Ormond Street Hospital, noted the effectiveness of thermally efficient internal blinds (with solar transmission <7%) in this mixed-mode building. The blinds reduced the operative temperature (which combines both air and radiant temperature) simply by cutting down direct radiation from the windows.

In fact internal blinds appeared to have a greater effect than external shading in this case, though it was recognised that as their effectiveness is dependent on human control, external devices would be a more reliable way of minimising solar gain. Internal blinds are very easy to retrofit, offering a straightforward means of upgrading an existing building. However, heat may build up between the glazing and the blind if that space is not ventilated, finding its way into the room by convection and raising the air temperature.

4.11 Phase 2 of the redevelopment of Great Ormond Street Hospital, London, designed by Llewelyn Davies Yeang.

● ● ●
CONDUCTION

Heat gain through solid elements – what's the right level of insulation?

A warming climate adds a level of complexity to establishing how much insulation will be appropriate for a project, bringing the need to consider overheating in summer as well as minimising energy consumption in winter. Conclusions as to the "right" level of insulation differed markedly across the Design for Future Climate projects.

In winter, the role of insulation is straightforward. It reduces heat loss through the building fabric and retains useful incidental gains more effectively. In summer, the situation is more complicated. Where external temperatures are close to internal ones, insulation is irrelevant. In more extreme high temperatures, it can help to isolate the interior from adverse external conditions.

There is generally considerably less difference between a comfortable internal temperature and the external *air* temperature in summer than in winter. However, when solar radiation is taken into account, a building's external surface temperature can be raised significantly, with a corresponding increase in heat transfer by conduction.

To size cooling systems, engineers use a calculated hypothetical outside surface temperature (the sol-air temperature) that takes into account the effects of radiation, which gives a useful indication of how much the external surface temperature of a building can be raised by the sun. The graphs of sol-air temperature for the July baseline climate in the south-east show that for dark-coloured roofs and east-, south- and west-facing walls, peak temperatures can increase to around 55°C, more than 20°C above a comfortable indoor temperature. A light-coloured roof might rise to around 35°C (see figure 4.12 overleaf).

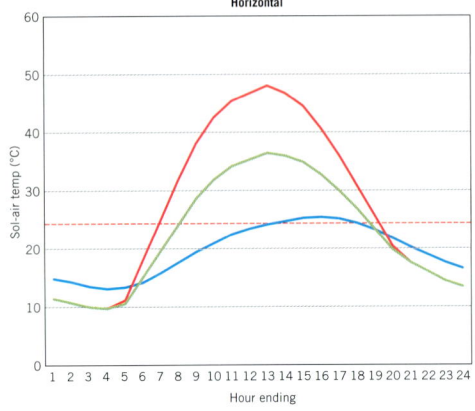

Horizontal

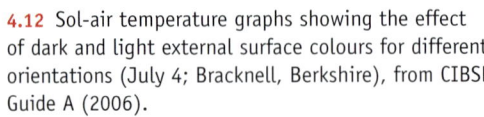

4.12 Sol-air temperature graphs showing the effect of dark and light external surface colours for different orientations (July 4; Bracknell, Berkshire), from CIBSE Guide A (2006).

— Air temp.
— Dark
— Light
---- Internal temp.

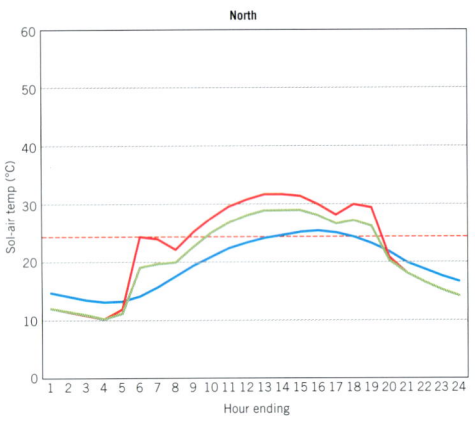

North

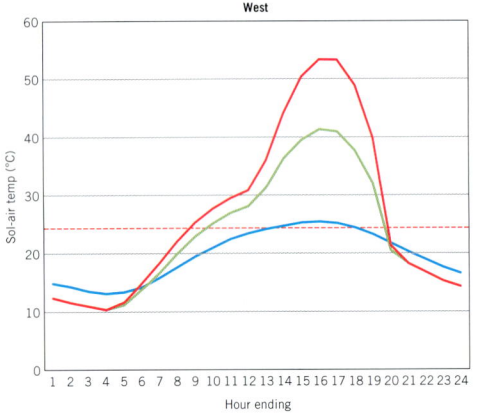

West

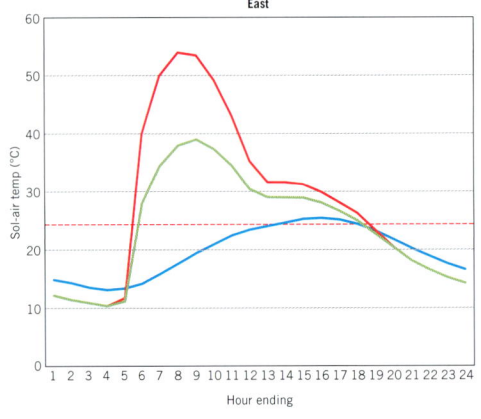

East

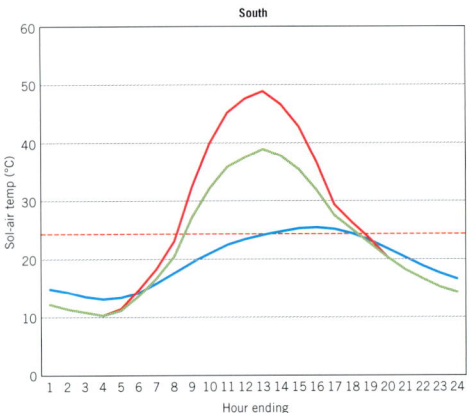

South

The Church View team were inspired by the widespread vernacular practice in warmer countries of painting buildings in light colours to achieve a high albedo (the ability of a surface to reflect radiation). They tried to investigate ways of using this technique, but found that determining values for the variables was not simple – and that, in any event, there was limited potential application in the context of a listed building. They also attempted to examine how dark-coloured external parking surfaces around the building affected the temperature of the air entering the building, but this was not possible with the dynamic simulation software they used.

It is worth noting from the sol-air-temperature graphs that the periods of high summertime temperature differences tend to be relatively short compared with general winter temperature differences. For most well-insulated new buildings, heat gains through solid building fabric will be of an order of magnitude less than those due to solar gain through windows or internal gains, but further research to establish the magnitude of positive effects of high-albedo finishes would be worth pursuing.

Gale & Snowden demonstrated that high levels of insulation, as part of an integrated Passivhaus approach for the Extra Care 4 Exeter project, would help cope with extreme conditions later in the century. Their strategy depended on minimising all unwanted gains (including hot air from outside and heat conducted through the building fabric) so that the "coolth" stored in the building fabric can maintain comfortable temperatures until external temperatures drop to a point where windows can be opened again.

On the other hand, Triangle Architects, working on a similar facility – the British Trimmings extra-care scheme, at Leek in Staffordshire – found that increasing insulation levels to Passivhaus standards would actually increase overheating when used with lightweight construction, and only made a marginal difference with heavyweight construction.

4.13 The British Trimmings extra-care scheme, Leek, Staffordshire, designed by Triangle Architects.

It is difficult to draw conclusions from these very different views without going into their assumptions in detail. However, one possible factor is that the Leek extra-care scheme did not have the benefit of cross ventilation (and thus was less able to exploit night cooling), and that its team might have assumed higher incidental gains. These would contribute usefully to heating in winter, reducing the advantage of higher insulation, but lead to overheating in summer as high insulation levels would prevent heat escaping through the walls when external temperatures dropped. The interactions between different elements of a strategy are complex and must be evaluated within the particular circumstances of a project.

For buildings with high internal or solar gains, the most pressing concern is to get rid of those gains. Under these circumstances, a number of teams found high levels of insulation to be counter-productive.

Perhaps the most extreme case was noted by the BRE team, working on a new Cardiff headquarters for Admiral Insurance. This deep-plan comfort-cooled densely occupied multi-storey office block has such a high level of internal gains that lowering the wall U-values (from $0.28W/m^2K$) increased demand for cooling, and had only a minimal effect on the energy used for heating. This building is also highly glazed, and a similar effect occurred when lower U-value glazing was substituted for the $1.76W/m^2K$ units specified.

In fact, the team's opinion was that if they had been able to build to a lower standard of insulation than that required to meet current Part L regulations, the building would have used less energy overall. They did, however, acknowledge that this was a direct result of needing to get rid of "waste" heat from internal gains and that, if more energy-efficient equipment became available, the situation was likely to change.

Determining the appropriate level of insulation for a project requires striking a balance between heat gains and losses throughout the year, which will depend not just on environmental conditions but also on the pattern and intensity of the building's use and the energy consumption of its equipment and services – all of which are uncertain. Regulations need to acknowledge this complexity, so that they steer designers towards providing buildings that are comfortable all year round and do not inadvertently preclude the best balance of overall energy consumption and carbon emissions. And, of course, they must do this in as simple and transparent a way as possible.

Green roofs

The multiple benefits of green roofs were promoted across a number of projects: providing useful amenity space, attenuating rainfall (see Chapter 7: Water), helping to reduce the urban heat island effect, improving biodiversity by providing vegetated areas, and reducing heat gain through roofs by a combination of thermal mass, a degree of insulation and evapotranspiration.

However, a definitive, quantified evaluation of the impact of roof-greening on heat transfer is not readily available, and teams struggled to establish its effectiveness. In many cases, green roofs seem to have been proposed on the basis of their general benefit without a very clear understanding of how they would actually improve internal comfort.

Gale & Snowden worked with Exeter University to model the role a green roof could play in keeping the single-storey dining room of their extra-care scheme cool. The results were somewhat inconclusive, owing to the limitations of dynamic simulation software and the complexity of the heat-transfer mechanisms involved. However, the team did draw the following conclusions.

- Green roofs are a more effective way of using vegetation to cool buildings than external planted areas because they are in direct contact with the building.
- The cooling effect on the room below was estimated to be between 0.5°C and 8°C – but probably closer to the lower figure as evapotranspiration takes place at the surface of the soil, remote from the building skin.
- The effectiveness of cooling is highly dependent on wind speed (high wind speeds reduce the cooling effect, by removing the cooled air produced by transpiration at the surface of the grass roof) and on keeping the soil damp.

This last point is particularly important because the cooling effect of evapotranspiration will no longer be available if the soil dries out – an increasing likelihood in future, given projections of drier and warmer summers. Irrigation is an option, but, given the probability of water shortage in many areas of the country, we must ask whether this is a sensible use of a limited resource. This is covered further in Chapter 7.

Clearly more research is needed so that when considering green roofs, designers can make realistic assumptions about their cooling potential as well as any other benefits.

● ● ●
HOT AIR

We now put considerable effort into making our buildings airtight for winter and into controlling ventilation rates to minimise the amount of cold outside air that needs to be heated to provide us with fresh air to breathe, and to remove odours. Mechanical ventilation with heat recovery (MVHR) systems, which provide a small but continuous supply of fresh air that is prewarmed by outgoing ventilation air, are becoming more commonplace as a means of achieving this.

Outside the heating season, ventilation is one of the principal means of getting rid of unwanted internal heat gains. One of the key design drivers for naturally ventilated buildings is the need to maximise their ability to ventilate effectively.

In the current UK climate, even in midsummer, external temperatures are generally lower than acceptable internal comfort temperatures. Even when they are not, the cooling effect of air movement has generally compensated for the fact that ventilation air is warmer than internal conditions. This is the basis of our intuitive response of increasing ventilation as a means to reduce overheating – the more the better. However, as a number of teams noted, with higher

MVHR provides good air quality in bedrooms at night when windows are shut

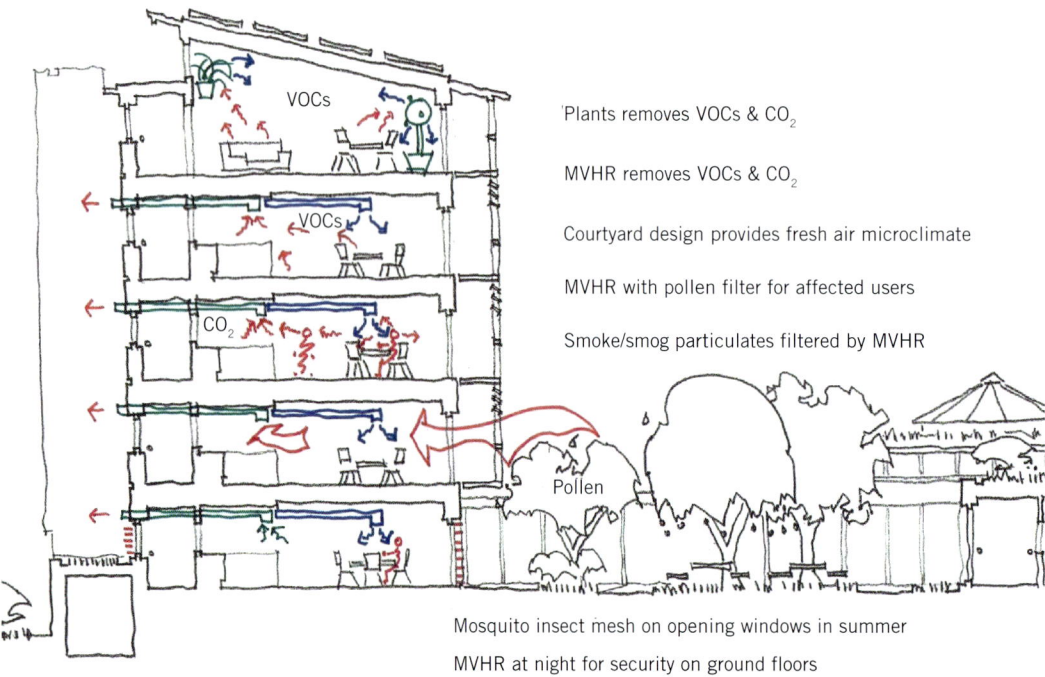

Plants removes VOCs & CO_2

MVHR removes VOCs & CO_2

Courtyard design provides fresh air microclimate

MVHR with pollen filter for affected users

Smoke/smog particulates filtered by MVHR

Mosquito insect mesh on opening windows in summer

MVHR at night for security on ground floors

Building and Landscape design working together to provide healthy environments

4.14 A summary of Gale & Snowden's air-quality strategy for the Extra Care 4 Exeter project.

summer temperatures will come a corresponding rise in the amount of time when external temperatures are higher than internal ones. When this is the case, high levels of ventilation will only exacerbate an overheating problem, and it would be better to switch to a minimal ventilation regime to minimise heat gain.

This is an important component of the strategy adopted by Gale & Snowden on the Extra Care 4 Exeter project in order to deal with high temperatures later in the century. They suggested that, by 2080, to stay comfortable it will be necessary to switch to a minimum-ventilation regime using the MVHR system (in summer bypass mode) to keep hot external air out of the flats. Importantly, this strategy presupposes that internal gains are minimised (as discussed on page 59). They proposed automating the switchover – automatically shutting windows and alerting staff to the change. But they also recognised that this is likely to appear counter-intuitive, particularly to the vulnerable client group living in the building, and that considerable support would be required from staff, who would themselves need to be trained.

A further benefit of the MVHR system noted by the team was that it could be used to filter out pollen and particulates, as well as the increased levels of volatile organic compounds (VOCs) likely to result from high temperatures.

5 DESIGNING FOR THERMAL COMFORT: Cooling Strategies

In the UK, it should be possible to design most buildings to be free running in summer, ie able to remain comfortable without the aid of mechanical systems. Rising summer temperatures will make this increasingly difficult, but the findings from the Design for Future Climate projects provide useful pointers on how to delay or avoid the need for mechanical systems as the climate changes.

VENTILATION

In a free-running building, natural ventilation is the principal means of removing heat. It is therefore important that the building is designed to promote and control effective purge ventilation. This requires much larger openings than those needed to provide the minimum fresh-air levels on which designers currently tend to focus, and it can have a significant impact on the elevational treatment of the building. We have traditionally relied on simple windows as the principal means of providing ventilation. However, to be fully effective a system should be able to provide the required amount of air at times, or under environmental conditions, when windows have shortcomings, providing ventilation that is:

- secure – at night, and when the building is unoccupied
- unrestricted – by blinds or guards to prevent people or objects falling out
- weather tolerant – when it is raining, etc

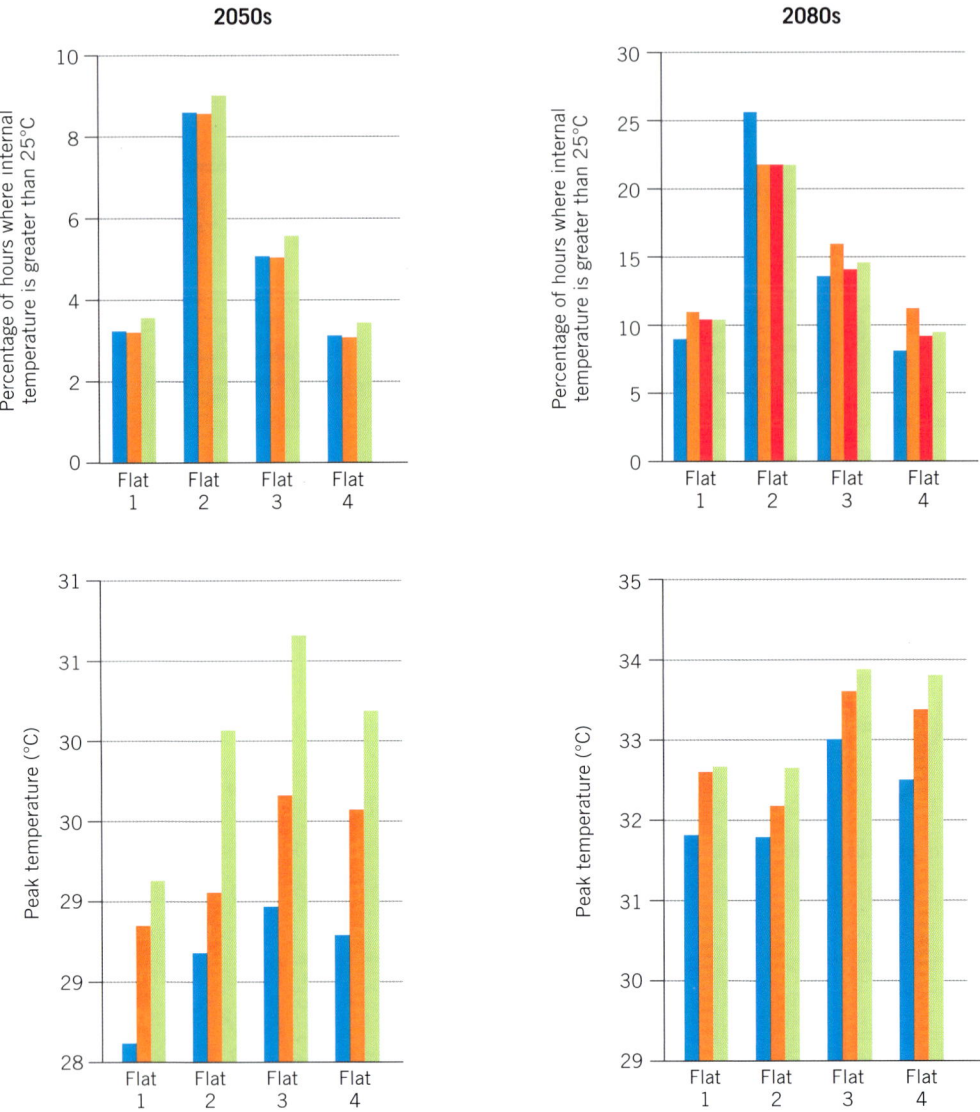

Key

Flat 1 – full cross-flow ventilation with opeing-restricted duct and windows

Flat 2 – single-sided ventilation

Flat 3 – full cross-flow ventilation with restricted night ventilation

Flat 4 – full cross-flow ventilation with opening windows either side

■ Heavyweight construction (Passivhaus standard)

■ Medium-weight construction (Passivhaus standard)

■ Lightweight construction (Passivhaus standard)

■ Medium-weight construction (2006 Building regulations standard)

5.3 Gale & Snowden modelled the impact on overheating of different combinations of ventilation strategy and construction type for the Extra Care 4 Exeter project.

5.4 A sectional view of the "rationalised design" for the British Trimmings extra-care scheme.

On the British Trimmings extra-care scheme, Triangle Architects investigated a number of design options, from an "adapted design" that remained within the parameters of the existing planning permission, to more experimental approaches involving replacement facades (a "rationalised design"). The adapted design used "air chimneys" to overcome the shortcomings of the conventional corridor-based extra-care scheme layout, which allowed only single-sided ventilation. They also introduced openings between floor levels as a means of ventilating the corridors themselves, prone to overheating because they contain services such as hot-water distribution pipework and lighting. This intervention had a significant knock-on effect on the fire-compartmentation and smoke-control strategy.

The different responses to the design of these two extra-care projects illustrate the impact of future climate on standard design solutions. Gale & Snowden chose to fundamentally redesign the conventional archetype to provide cross ventilation, principally through windows. Triangle Architects chose to adapt it by finding vertical routes through the building to improve cross ventilation. The former is a significant departure that fundamentally changes the way schemes like these would operate; the latter sticks more closely to the tried-and-tested archetype, but does mean building in "dead" space (the ventilation chimneys) and necessitates a more complex fire-compartmentation regime. Both approaches are valid, deriving from the particular circumstances of design team, client and site. However, by being allowed to change the fundamentals of the scheme Gale & Snowden were apparently able to provide an adapted design at no significant extra cost.

One issue, again noted by Gale & Snowden, was that the effectiveness of ventilation as a means of reducing summertime temperatures differed significantly between modelling carried out in IES and in PHPP (Passive House Planning Package, the standard spreadsheet used to validate Passivhaus designs). They concluded that dynamic simulation software like IES may be "too perfect" in modelling the effect of opening windows in line with the parameters input into the model (eg the windows open by 50% when the temperature exceeds 25°C). In reality, if window opening is controlled by people they tend not to react immediately and do not necessarily open the windows as much as required. It would appear that PHPP takes a rather more pragmatic view of what happens in practice, and therefore downplays the effectiveness of natural ventilation.

Dynamic simulation models are an essential tool for predicting the behaviour of buildings under summertime conditions, but care needs to be taken to ensure that the assumptions made in the model are delivered by the building in practice. Buildings that rely on natural ventilation to maintain comfortable conditions depend on the integrated operation of many separate components – some automated, some manually operated; some specified by the architect, some by the environmental engineer or by other specialists, such as acoustic engineers. This integration is easier to achieve in a model than in practice, and there is a real danger that the gap between theory and real life will amount to the difference between success and failure.

AEDAS, working on the Harris Academy in Purley, south London, carried out detailed analysis of the aerodynamic, free ventilation area through the windows as designed, ie once any obstructions and the width and shape of the openings have been taken into account. They noted that the configuration of the window, its opening gear and its position in a reveal all impacted significantly on the free area of that window when open, and therefore on the effectiveness of the ventilation strategy. Similarly, they highlighted the relationship between the free area required through the external facade and that required to provide cross ventilation for the classrooms via internal, acoustic transfer grills. These components were procured as separate contract packages and specified by different consultants, and it is essential that the environmental-design link between them does not become disconnected during the detailed design phase.

AEDAS also noted the potential for high-level automated openings to be blocked by window blinds, and tested the robustness of the automated ventilation strategy by rerunning the model with half of the openings blocked. The effect on overheating was dramatic, more than doubling the hours when temperatures topped 28°C. These are all commonly occurring issues in the field, and are difficult to solve without adding considerable complexity – the enemy of robust design. If they are not solved in practice, however, any theoretical summer comfort strategy is invalid.

A number of teams noted the difficulty of providing secure ventilation, particularly where buildings face onto public or insecure areas. Again, this is an essential component of a successful night-ventilation strategy. If it cannot be operated reliably in practice, the entire strategy may be invalidated.

Similarly, the sizeable opening areas required for summer ventilation can be difficult to achieve in a noisy environment – the need to attenuate the ventilation path can result in quite large enclosures that are challenging to accommodate architecturally. Of course noise may only be a temporary problem, as noted by Hoare Lea working on the University of Greenwich project, where a facade was sealed against noise from trains in a nearby cutting. The designers suggested that if, in future, the cutting was enclosed there would be significant benefits in providing opening windows on this facade. Arup, working on 100 City Road near the busy Old Street roundabout, made a similar point, anticipating a time when electric vehicles would be the norm and heavy road transport reduced.

Interestingly, conclusions on the impact of rising summer temperatures on mixed-mode buildings also differed. At the Oxford University Press offices, Hoare Lea found that the use of mechanical ventilation and cooling was likely to increase from 54% in the control year to 70% in the 2080s (COPSE files, Medium emissions), with a corresponding decrease in the period during which the building could operate with only natural ventilation (see below).

But WSP Built Ecology, working on Great Ormond Street Hospital, came to a different, counter-intuitive, conclusion. They found that milder winter conditions would increase the number of hours when the outside temperature was warm enough to open the windows (ie over 12°C) and that, up to the 2050s, this was greater than the reduction in the number of hours in summer when external temperatures were too high for natural ventilation to be effective.

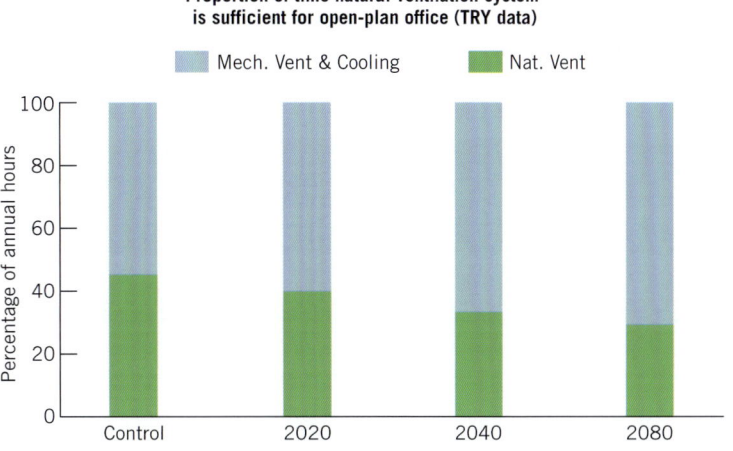

5.5 At the Oxford University Press offices, Hoare Lea found that mechanical cooling would be required for a greater percentage of time later in the century.

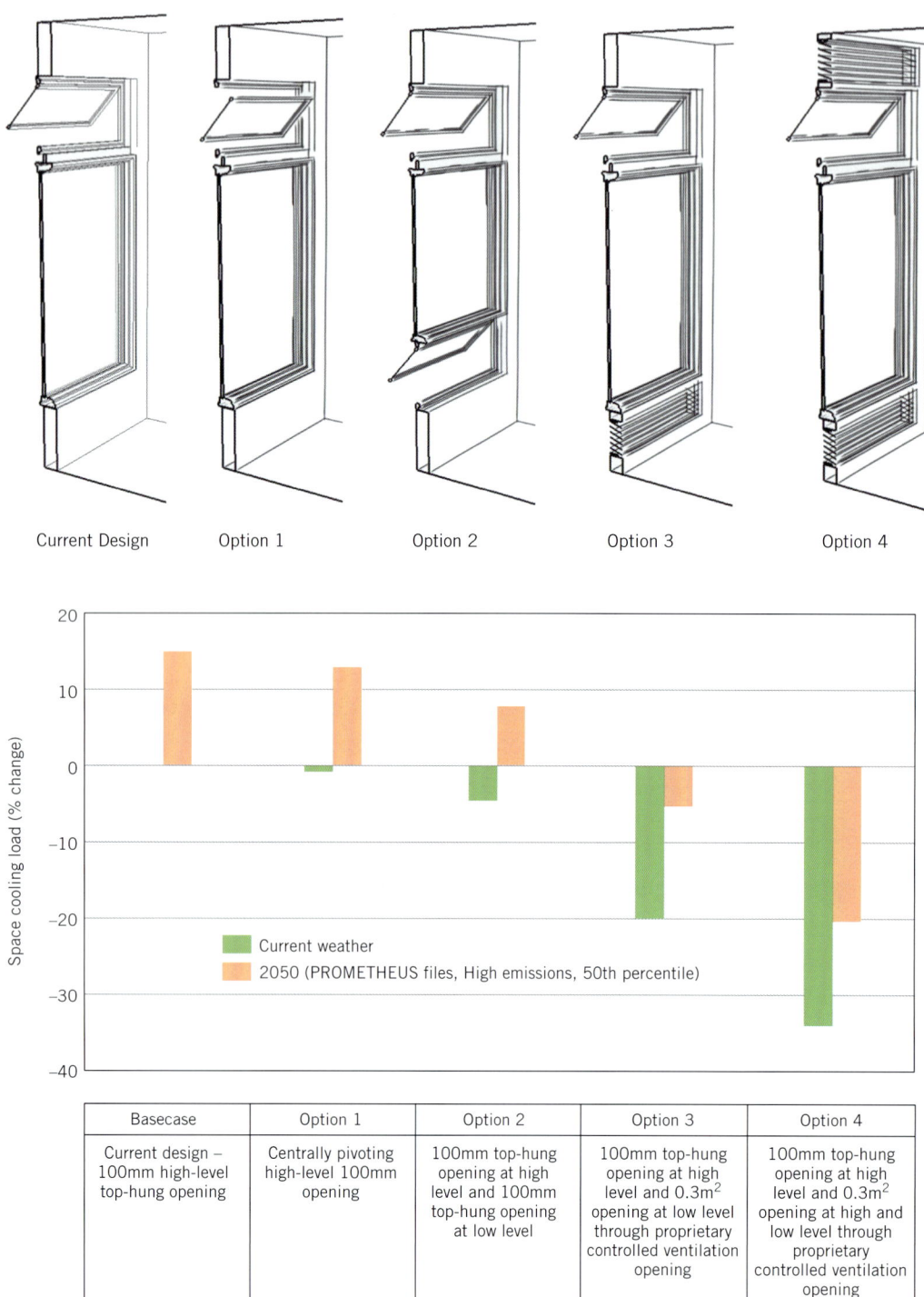

| | Current Design | Option 1 | Option 2 | Option 3 | Option 4 |

	Basecase	Option 1	Option 2	Option 3	Option 4
	Current design – 100mm high-level top-hung opening	Centrally pivoting high-level 100mm opening	100mm top-hung opening at high level and 100mm top-hung opening at low level	100mm top-hung opening at high level and 0.3m^2 opening at low level through proprietary controlled ventilation opening	100mm top-hung opening at high level and 0.3m^2 opening at high and low level through proprietary controlled ventilation opening

5.6 WSP Built Ecology modelled a number of window designs (shown above) to maximise natural ventilation at Great Ormond Street Hospital, and produced this graph to show how they would affect demand for space cooling.

The disparity between these two conclusions may result from a number of different factors: location, orientation, construction and hours of use. Again, it demonstrates the need to consider each intervention on its merits – and not let preconceptions exclude counter-intuitive options.

To maximise the effectiveness of natural ventilation, the team explored a range of window-design options with different openable areas and opening configurations (see figure 5.6 opposite).

Their analysis of the effect on the cooling load showed that there was a clear relationship between the size of opening and its effectiveness, and that a combination of high- and low-level window openings is more effective than a single opening. The most effective configuration (Option 4) showed a reduction in cooling demand of more than 30%, which could more than mitigate the predicted increase in space cooling loads up to 2050.

This finding is particularly encouraging because it shows an ongoing role for natural ventilation – even for a building in an urban heat island, where air quality is paramount and which is operational 24 hours a day.

When setting building plan depths, it is common to use the rule of thumb that cross ventilation is most effective to a depth of approximately five times the floor-to-ceiling height. This rule was explored in reverse for 100 City Road in London, a 40m-deep mixed-mode office building designed by AHMM with engineers Arup, with a proposed floor-to-ceiling height of 3.5m. Their hypothesis was that increasing the floor-to-ceiling height might improve cross ventilation, enabling the building to operate in natural-ventilation mode for longer and reducing use of the proposed mechanical-ventilation and chilled-ceiling system. They additionally anticipated that higher ceilings could also result in greater stratification of air at different temperatures, with warm air rising above the occupants and reducing the cooling load, but that this might be negated by some mixing of the air due to the chilled ceilings.

Arup used computational fluid dynamics (CFD) software to model the changes in air distribution and movement for three floor-to-ceiling heights: 3m, 3.5m and 5m. This showed that the chilled ceiling (with a surface temperature of approximately 20.5°C) did not, in fact, prevent stratification for any of the room heights modelled. Increasing the floor-to-ceiling height did improve stratification, but average space temperature rose due to the additional solar gains through the increased area of glazed facade.

Working on the University of Sheffield Graduate School, Arup found that increasing the floor-to-ceiling height from 3.6m to 4.8m (without a corresponding increase in glazed area) reduced the number of overheating hours in two of the spaces they studied by around 20%, though they acknowledged the cost and planning impacts of such a significant change.

On City Road, Arup also used CFD analysis to compare the effect of high- and low-level mechanical ventilation strategies on the distribution of air temperature in the space (see figure 5.7 below). The chilled ceiling was used in conjunction with the high-level option, but not for the low-level (displacement) system, where all cooling was to be provided by the under-floor air supply, requiring a higher ventilation rate.

They found that the average air temperatures in the space were essentially the same for both options but that, with the displacement system, the air was more stratified, keeping the air temperature in the occupied zone 1-2°C lower than for the high-level option.

They suggested that the high-level supply option had some advantages in offering greater adaptive capacity. The level of cooling the system provides can be increased by reducing the air supply temperature or increasing the ventilation rate without causing unacceptably low temperatures or high air speeds in the occupied zone. On the other hand, they pointed out that increasing the floor-to-ceiling height would increase the efficiency of the displacement option as this would allow stronger stratification to develop.

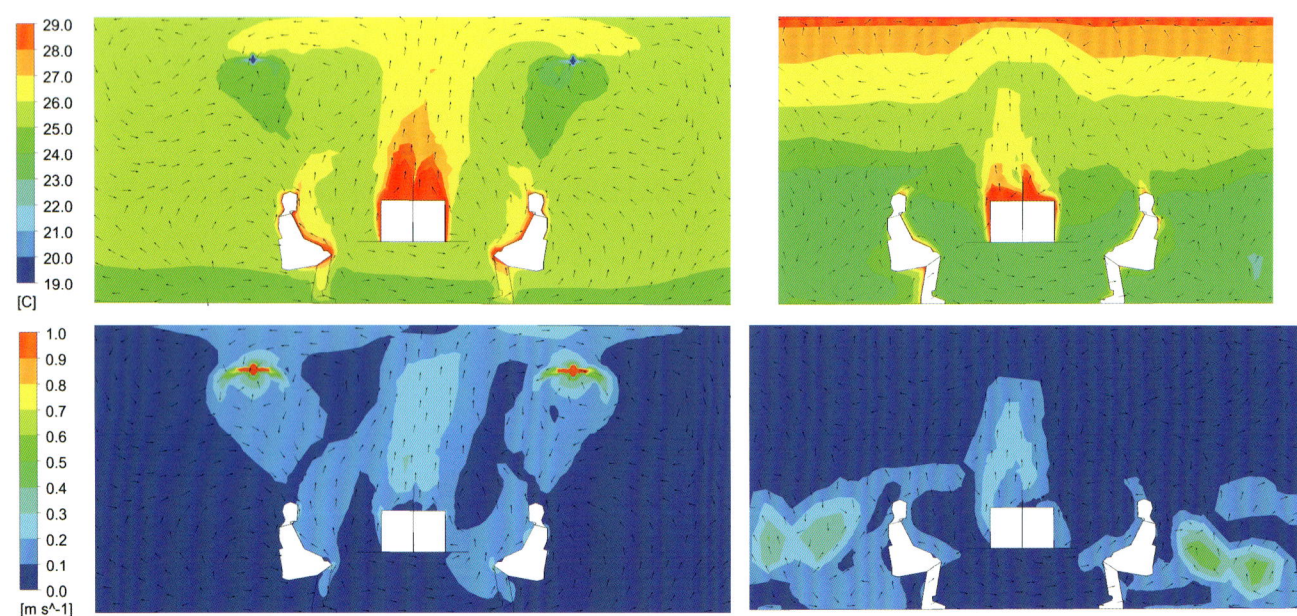

5.7 At 100 City Road, Arup used CFD modelling to assess air movement and temperatures for different types of ventilation and ceiling heights. These graphs show air temperature (top row) and velocity (bottom row) for a space with a 3m ceiling height, for high-level ventilation (left) and for low-level displacement ventilation (right).

Using the ground to pre-cool ventilation air

Increasing external air temperatures will reduce the cooling effect of ventilation. In response, the teams explored low-energy opportunities to pre-cool the air used for ventilation. The ground is a large heat sink, and below around 2m it remains at a constant temperature, equal to the average annual temperature. This means that in summer it is cooler than the air temperature at the surface, and in winter, warmer. Closer to the surface, the temperature fluctuates – more nearer the surface, less lower down – lagging behind the seasons through the dampening effect of the thermal mass of the soil.

Gale & Snowden explored ways of using the ground as a source of coolth to provide pre-cooled air to the MVHR system when the building was in minimal ventilation mode during hot spells, either through an underground air duct or a piped heat exchanger similar to those used for ground-source heat pumps. Underground air ducts require very long lengths in order to be effective, and there was not enough space on the site to accommodate them. The manufacturer's technical information for the piped system indicated that ventilation air temperatures could be reduced by up to 15°C; for the purpose of analysis, Gale & Snowden used more conservative estimates of 5°C and 10°C.

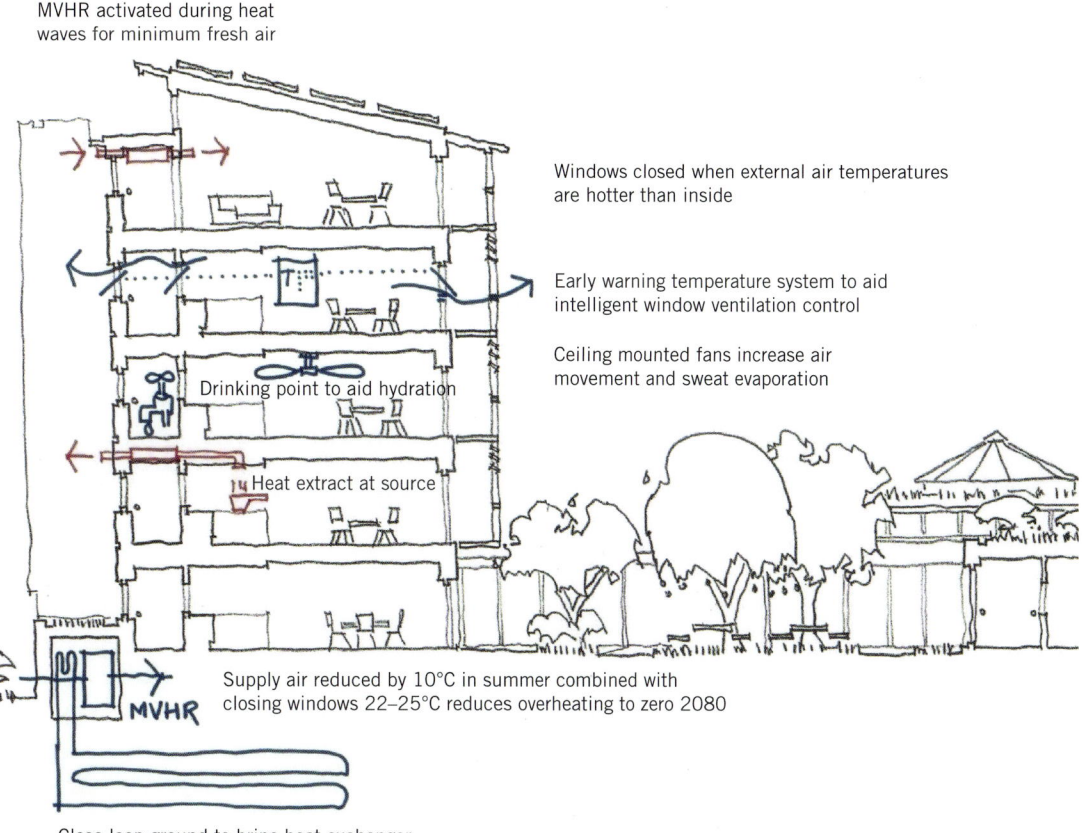

MVHR activated during heat waves for minimum fresh air

Windows closed when external air temperatures are hotter than inside

Early warning temperature system to aid intelligent window ventilation control

Ceiling mounted fans increase air movement and sweat evaporation

Drinking point to aid hydration

Heat extract at source

Supply air reduced by 10°C in summer combined with closing windows 22–25°C reduces overheating to zero 2080

MVHR

Close loop ground to brine heat exchanger

5.8 This diagram shows the active adaptation strategies considered by Gale & Snowden for the Extra Care 4 Exeter project, including the underground cooling system.

With a lower supply-air temperature, there was a dramatic effect on overheating hours, but the cost of accommodating the system was significant. This was due not only to the groundworks but to the multi-storey nature of the building and the need to ensure fire compartmentation between different areas. It was noted, however, that the principle could be applied more easily to building types which did not require extensive zoning or fire compartmentation, such as open-plan offices and schools. Such systems must be designed in from the start so that they can be simplified as far as possible, and there needs to be sufficient ground area on site. The piped system has greater potential for widespread application than the underground air-duct approach, because it requires less ground space and there are no issues of air contamination (from bacteria potentially developing in the underground ducts).

Triangle Architects, working on the British Trimmings extra-care scheme, also suggested the use of ground-linked air ducts to pre-cool ventilation air as one component of an ingenious replacement facade, combining pre-cooled air with solar-driven extraction (see figure 5.9 opposite). This was designed to suit a hot–humid climate, which they identified as a possibility in the most extreme 2080 projections. Modelling of this system showed a dramatic improvement in comfort, but solar-driven ventilation is a rarity in the UK, and in practice it has not always been as successful as predicted. For this approach to gain traction, it would be necessary to test the modelling against reality in order to validate modelling algorithms and assumptions.

The team also noted the potential for using ground-source heat pumps as an alternative to standard gas boilers to provide underfloor cooling as well as winter heating, though the effects of this strategy were not evaluated in any detail. In many ways, this is a better use of ground-source systems than if they are used only for heating: the heat removed from the ground in winter is balanced to some extent by heat returned to it in summer.

5.9 On the British Trimmings project, Triangle Architects explored the idea of a replacement facade combining pre-cooled air from the ground with solar-driven extraction.

Use of planting and landscape to pre-cool ventilation air

A number of teams noted the potential for external landscaping to reduce air temperatures around buildings, and thus enhance the cooling potential of natural ventilation.

Some teams referred in passing to the use of water features and water bodies, though without presenting any design data. There were similar general comments on the role of landscape and planting from many teams, and here a limited number of projects did attempt to quantify the possible benefits.

An investigation carried out by the University of Exeter for the Extra Care 4 Exeter scheme offers a useful summary of available research on the effect of planting, which provided evidence that planting would modify outdoor conditions and would have some impact on conditions inside the building.[9] This was due to:

1. increased shading on windows and the building surface
2. reductions in air temperature outside the building from changes in albedo and evaporation (including transpiration)
3. reductions in light being reflected into the building
4. the reduced external temperature changing the flow of air through the building via any stack effects
5. reductions in wind speed from increased surface resistance.

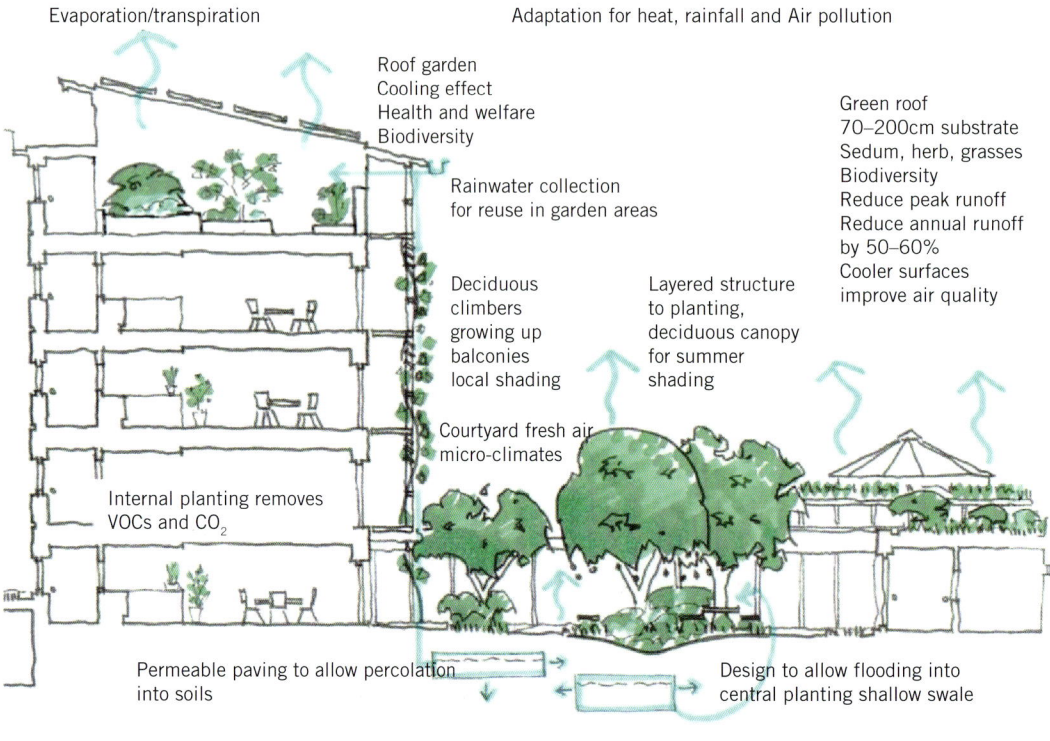

Evaporation/transpiration

Adaptation for heat, rainfall and Air pollution

Roof garden
Cooling effect
Health and welfare
Biodiversity

Green roof
70–200cm substrate
Sedum, herb, grasses
Biodiversity
Reduce peak runoff
Reduce annual runoff
by 50–60%
Cooler surfaces
improve air quality

Rainwater collection
for reuse in garden areas

Deciduous
climbers
growing up
balconies
local shading

Layered structure
to planting,
deciduous canopy
for summer
shading

Courtyard fresh air
micro-climates

Internal planting removes
VOCs and CO_2

Permeable paving to allow percolation
into soils

Design to allow flooding into
central planting shallow swale

Sequence of rainwater storage crates for natural
percolation to planting and pumped irrigation

Green micro-climate reduces summer
temperatures by 3°C

5.10 A diagram of the landscaping strategy developed by Gale & Snowden for the Extra Care 4 Exeter project, showing adaptation measures for heat, rainfall and air pollution.

The investigation measured the changing external temperature from differences in albedo and evaporation using published empirical data on the impact of planting around buildings. As a result, it recommended that planting be seen as an adaptation strategy, but that additional modelling should take place using an accurate model of the building, and possibly "more sophisticated methods". It also noted that while there have been many studies of the impact of green spaces on urban environments and the urban heat island, there have been few on the impact of small-scale planting around individual buildings – a deficit that needs to be addressed.

Using a heavily simplified IES model, the report found that for both single-sided and cross-ventilated rooms, external planting did reduce internal temperatures at the hottest times of the year. The modest reduction in external temperature outweighed the potentially negative effect of an assumed reduction in wind speed. "A reduction of approximately 0.5°C in both mean and maximum internal temperature is seen with each 1°C of cooling provided by the plants." More research is clearly required into these effects.

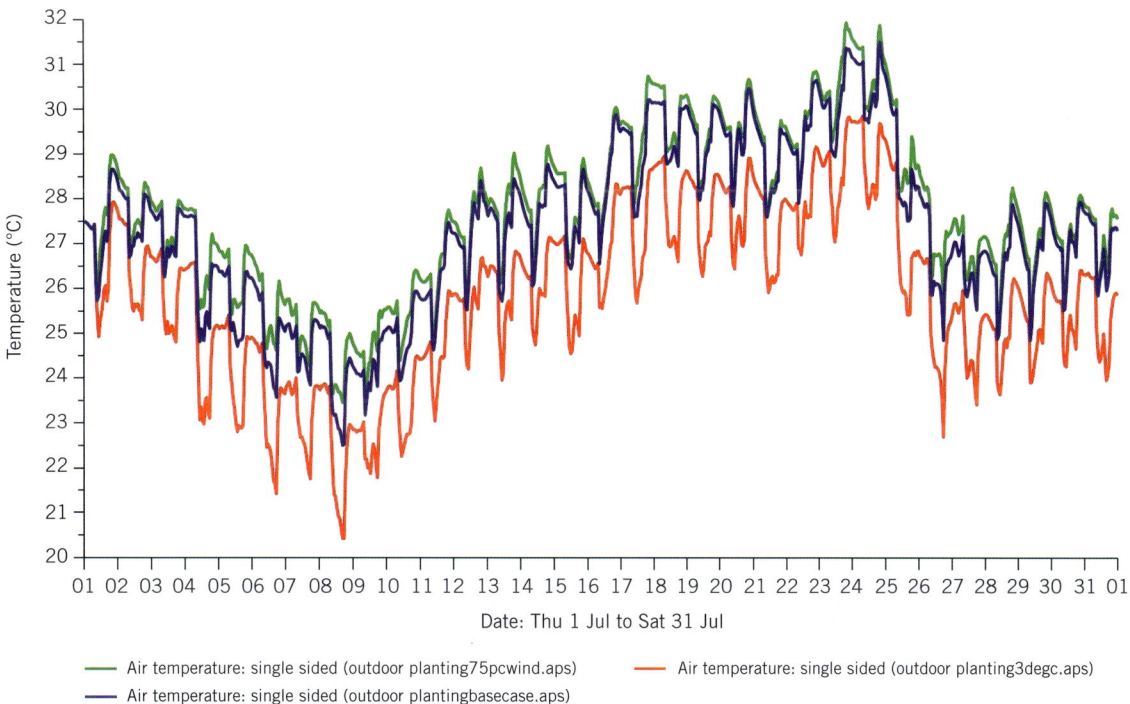

5.11 This graph was produced by the University of Exeter for the Extra Care 4 Exeter project, and shows the potential cooling effect of outdoor planting on the building. It shows internal temperatures in the month of July, using the current Test Reference Year for Plymouth.

The project also considered the potential for using indoor plants to provide additional cooling via evaporation. It concluded that the impact would be small, and suggested caution in using plants to improve thermal conditions. Each plant could supply a maximum of 2.6–13W of free cooling, but any positive effect on thermal comfort is likely to be offset by higher humidity during summer if there is insufficient ventilation.

Right at the start of the Church View project, Bauman Lyons consulted German landscape architect Professor Peter Latz, who has been applying landscape cooling techniques since the 1970s. His rule of thumb is that planting can make external spaces 2–3°C cooler. However, the majority of his experience relates to relatively large areas, and no data seems to have been made available for smaller-scale applications.

Green walls

Green walls were referred to as a potential source of cooling by a number of the projects. However, while they offer an undoubted (though difficult to quantify) benefit by shading a facade, there is little hard data on which to base an analysis of their potential to reduce the temperature of air entering a building. Also, while plants growing directly in the ground do not necessarily require irrigation, those in high-level planters do, bringing additional cost and maintenance as well as raising similar concerns to those associated with green roofs as to whether this is an efficient way of using water for cooling.

The Church View team referred to a report carried out by Stuart Archer at the University of Sheffield on the cooling potential of green walls. This noted that while thermal insulation and cooling are among the many benefits claimed by manufacturers of green walls, there is little or no data to back this up in the UK context. Experimental studies have generally been carried out in warmer climates such as Riyadh and Mumbai, with significant success. However, for cooler climates such as Sweden, green walls showed no effect for buildings insulated to good standards.[10]

Archer's report went on to model a simple one-storey building, looking at the effect of a green wall on insulated and uninsulated construction, for the contrasting climates of Nottingham and Dubai in winter and summer. For the insulated building, there were only small temperature differences, with a maximum cooling/insulating effect of 1°C or less. For the uninsulated building, the maximum cooling effect recorded was 2.4°C, in the Dubai summer.

The report also looked at water use, using weather data collected at the university. It found that the potential rainwater harvested from the roof was lower than the amount required to maintain the green wall, resulting in an overall water deficit where harvested rainwater would have to be supplemented with mains water.

5.12 The British Trimmings extra-care scheme
with a planted replacement facade.

The team on the British Trimmings extra-care project also studied the potential cooling benefits of a green facade for the hot–dry climate indicated by their projections for 2050 (PROMETHEUS files, Medium emissions, 50th percentile). This would cool the air entering the building using evaporative cooling through an irrigated planted facade, set 1m from the main facade. For modelling purposes, an evaporative cooling rate through transpiration of 1kg of water/hr/flat was assumed. This facade option appeared to be more effective than the ground-cooled air-supply option (described on page 92). However, if it relies on a continuous water supply at this rate, its consumption amounts to more than a quarter of the potable water target for a Level 5 or 6 Code for Sustainable Homes dwelling.

THERMAL MASS

As noted in the section on ventilation above, there is a close relationship between thermal mass and purge ventilation.

Thermal mass slows down swings in internal temperature by exploiting the time it takes for heat to be absorbed and released from a building's structure. It can therefore only contribute to keeping a building cool if the heat that has been absorbed during the day is removed overnight, so that heat absorption can begin again the next day when the building is occupied. This is usually achieved by ventilation with cool night air. Once cooled, a heavyweight building heats up slowly and peak temperatures are reduced, ideally remaining at a comfortable level until the end of the day – by which time the outside air has cooled down and the process can repeat.

There are situations where its effectiveness is limited. If a building is occupied continuously, there is no opportunity to pre-cool it and, therefore, little advantage in using heavyweight construction. Neither can the benefits of thermal mass be exploited if it is not possible to ventilate the building effectively, as in the examples given above. In fact, heavyweight construction can make matters worse, retaining heat that could be removed more quickly by ventilating a lighter-weight structure, as noted in the comparison of single-sided and cross-ventilated strategies for the Extra Care 4 Exeter project (see pages 82 and 84).

All the teams working on free-running naturally ventilated buildings used thermal mass in one form or another to reduce peak daytime temperatures.

As with ventilation, to be effective the levels of thermal mass that are included in a thermal model must actually be available in practice, ie in contact with both the space they are intended to cool and with sufficient, cool night ventilation. The Church View team decided that their model should exclude thermal mass in walls as they were likely to be covered by furniture, shelving and wall decorations. This limited available thermal mass principally to the ceiling, so they also explored the use of phase-change plasterboard to enhance its performance (see opposite).

For night ventilation to be effective, there must be a reasonable difference between night and daytime temperatures (diurnal swing). This should average at least 5°C, and preferably 7°C. Several teams also noted the impact of rising night-time temperatures on the effectiveness of night-time cooling – even when using the Adaptive Comfort Threshold methodology, which takes some account of our ability to adapt to generally rising temperatures as well as seasonal and daily variation.

 PASSIVE DESIGN ASSISTANT

Arup (working with AHMM architects and the Concrete Centre, and funded under a separate TSB programme) has developed a free software tool that can help designers get to grips with the relationship between ventilation and thermal mass, for both free-running passive buildings and those with mechanical systems. The Passive Design Assistant (PDA) can be used to inform early design decisions and, more generally, as an educational tool to communicate the principles of building physics to non-specialists. It is an excellent, approachable tool which allows designers quickly and easily to modify construction type, occupancy, ventilation opening size and timings, as well as location, equipment use, orientation and shading, giving an instant readout of the effects of any change. It can also be used to "sanity check" the results of more complex thermal-simulation models.

 PHASE-CHANGE MATERIALS

Phase-change materials offer a similar means of storing heat to conventional, thermally massive materials such as masonry or concrete, but are lighter-weight. They operate in a particular temperature band, typically around 25°C. As the temperature reaches this point, the material changes from a solid into a liquid, absorbing the additional latent heat that is required for it to change phase. This means that until a space reaches the melting point of the phase-change material, it behaves as a thermally lightweight environment. When that melting point is reached, the temperature then remains relatively constant while the material liquefies. Once all the material has changed phase, the space will revert to behaving like a lightweight environment. If it continues to gain heat, temperatures will rise more rapidly than in a conventional, thermally massive structure.

The Church View team found that dynamic simulation software was not able to model the specific effect of phase-change materials (although it is understood that plug-ins are now becoming available to do this). Instead, they simulated an approximation of the positive effect on reducing overheating by including thin (50mm) concrete walls in the model.

Thermal mass can also be beneficial in comfort-cooled and mixed-mode buildings. Removing false ceilings to expose thermal mass in the comfort-cooled new Admiral Insurance headquarters was shown to result in a moderate decrease in demand for cooling in 2080.

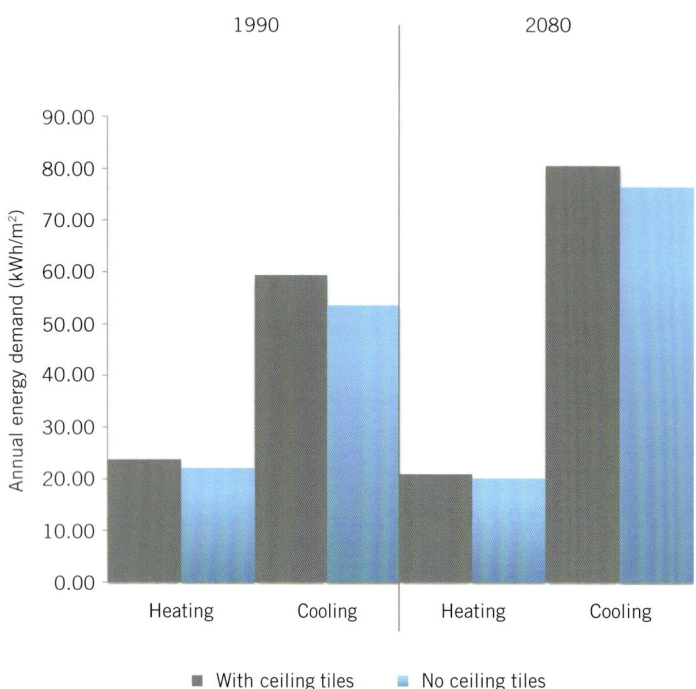

5.13 Increasing exposed thermal mass moderately reduced cooling demand in BRE's modelling for the new Admiral Insurance headquarters.

The team working on 100 City Road noted the dramatic effect of night ventilation for their mixed-mode building, reducing cooling requirements by 30% and temperatures by 2°C, even without the active cooling of the slab that was included in their baseline scheme. They also explored the effect of different concrete slab thicknesses, modelling 250mm, 300mm and 400mm options. Adoption of the thickest slab showed a small reduction in cooling load – just 2% in the current climate, and less than 0.5% for future weather files. However, this element of their research did not include the effect of night pre-cooling, so the reduction in cooling load is only that resulting from the effect of thermal lag. This reduction in cooling load was noted as negligible when compared with the additional embodied energy of the extra concrete.

Thermal mass may have embodied carbon impacts, which need to be taken into account when evaluating adaptation strategies that rely on it. It used to be reasonable to ignore embodied impacts on the assumption that, over a building's long lifetime, the energy consumed by its operation would be many times greater than the energy and carbon embodied in its structure. As buildings become more efficient this balance changes, and the embodied component of a building's carbon footprint becomes more significant. Commercial buildings, in particular, may have remarkably short lifetimes – perhaps only 20 or 30 years between rebuilds or major refurbishments – and it is becoming apparent that over such short periods there is a much finer balance between embodied energy and energy in use. We therefore need to consider whether a heavyweight approach with a high embodied energy fully "pays back" the initial carbon "investment" with the carbon savings it delivers in use.

The results of the research for 100 City Road demonstrate the importance of this issue, as does recent work by engineer Max Fordham which compared light and thermally massive options for a mixed-mode office building.[11] This showed that the embodied carbon of the heavyweight option was equivalent to more than 100 years of the annual saving in cooling that the thermal mass actually delivered. However, it should be noted that the building examined already required cooling and that this argument will not necessarily hold true for free-running buildings. Similarly, the study was based on a specific building and made assumptions about energy consumption that may not be delivered in practice.

Loss of available thermal mass was cited as one disadvantage of installing internal, rather than external or cavity, insulation when thermally upgrading existing buildings. This is of particular relevance for homes, given that upgrading the existing domestic stock is a substantial element of the UK's strategy to reduce carbon emissions. There are concerns that widespread overheating could be an unintended consequence of these interventions, whether internal or external.

None of the first tranche of Design for Future Climate projects focused on domestic refurbishment. However, a recent Master's thesis by Tessa Barraclough, studying at the Graduate School at the Centre for Alternative Technology, reached some interestingly counter-intuitive conclusions in an analysis of emerging data from some of the Retrofit for the Future projects (also funded by the TSB). She found that overheating was occurring already and seemed to be more prevalent where external insulation had been used, particularly in the case of bedrooms. While recognising the danger of extrapolating from a very limited sample, she postulated that where rooms are not sufficiently well-ventilated, higher thermal mass can mean that the heat that has built up during the day is retained for longer after the external air temperature has started to drop, making the room uncomfortable.

These issues again highlight the need to understand and explore the interplay between thermal mass, heat gain and ventilation for the particular uses and circumstances of a building – keeping a suitably pragmatic eye on how theory can be delivered in practice.

AIR MOVEMENT

The cooling effect of air movement, whether natural or from a fan, largely depends on its speed. Typical subjective reactions are as follows:

< 0.25 m/s	unnoticed
0.25 – 0.50	pleasant
0.50 – 1.00	awareness of air movement
1.00 – 1.50	draughty
> 1.50	annoyingly draughty[12]

However, reactions are dependent on air temperature too. In a heated room, some air movement (>0.1m/s) is needed to avoid stuffiness – 0.25m/s is the typical upper limit to avoid complaints about draughts. Under hot conditions, 1m/s is generally regarded as pleasant and speeds up to 1.5m/s are acceptable, though higher speeds also have negative consequences such as papers blowing about.

Dynamic simulation software generally assumes an airspeed of 0.25m/s but, as noted by the Church View and Worcestershire schools teams, it cannot take into account the cooling effect of increased airspeeds under hot conditions. A number of teams noted the potential for desk or ceiling fans to cool occupants, the Worcestershire schools team suggesting that an airspeed of 0.5m/s would provide about 2°C of cooling effect.

For a warm climate, it has been suggested that the limit of tolerable airspeeds might be extended to 2m/s with a potential cooling effect of about 6°C, based on 1°C of additional cooling effect for every 0.275m/s increase in airspeed above 0.2m/s.[13] At these speeds, paperweights might come into their own!

It should be noted that these are still very low speeds compared with what we are used to outside – all fall into the "light air" or force 1 category on the 12-point Beaufort scale of wind speed. This highlights the value of being able to use outside spaces, where airspeeds are likely to be higher, as a source of potential relief from uncomfortable indoor conditions (see discussion of external spaces on pages 105–107).

Desk fans are a simple, individually controllable option that needs no special design considerations. Ceiling fans require reasonable ceiling heights; the standard recommendation is a minimum of 2.3m to the underside of the fan. This may limit their safe use in many domestic-scale environments and illustrates one of the advantages of 2.5–2.7m ceiling heights, as recommended in the interim edition of the Greater London Authority's London Housing Design Guide.

It may be an obvious point, but fans only make people *feel* cooler – they do not reduce the temperature of an enclosed space. In fact they increase it as a product of the energy they use, which means that running a fan in an unoccupied space is counter-productive.

IMPACTS ON MECHANICAL BUILDING SYSTEMS

There are cases where mechanical cooling is required under current conditions – in deep-plan densely occupied buildings, for example – and, realistically, rising temperatures will increase its use. Even if cooling is required, designers should aim to minimise the cooling demand through intelligent passive design.

However, whereas a building's basic form and construction are fixed for its entire lifetime, the systems within it will be replaced at regular intervals. There is also a hierarchy within the mechanical plant itself: the fundamental architecture (ducts, pipes, cores, etc) is more fixed than components such as boilers, chillers, pumps and fans. The expected lifespan of different elements varies correspondingly – some will be replaced every 20 or so years, others only at major refits of the building. For the long term, the key is to provide enough space (and structural strength) in the right places to accommodate services, taking account of potentially large levels of uncertainty about the changing climate. Decisions about individual, regularly replaced components can be based on a shorter time-frame with relatively high levels of certainty.

Heating and cooling plant is sized to deal with peak demands. Therefore, in the context of a warming climate, heating systems must be designed to cope with conditions at the beginning of their lifespan, whereas cooling systems need to take account of uncertain conditions at the end of theirs: an area of inherent uncertainty.

Overall, total energy consumption is expected to drop as decreasing demand for heating is not fully compensated by the increasing demand for cooling. However, the shift from gas to electrical power (on top of any corresponding switch in the transport sector from fossil fuels to electricity) presents two challenges. It will increase the challenge of decarbonising the electricity supply because more renewable energy will be required, and it will also strain the capacity of the National Grid infrastructure itself. The effect is illustrated by these graphs prepared for the

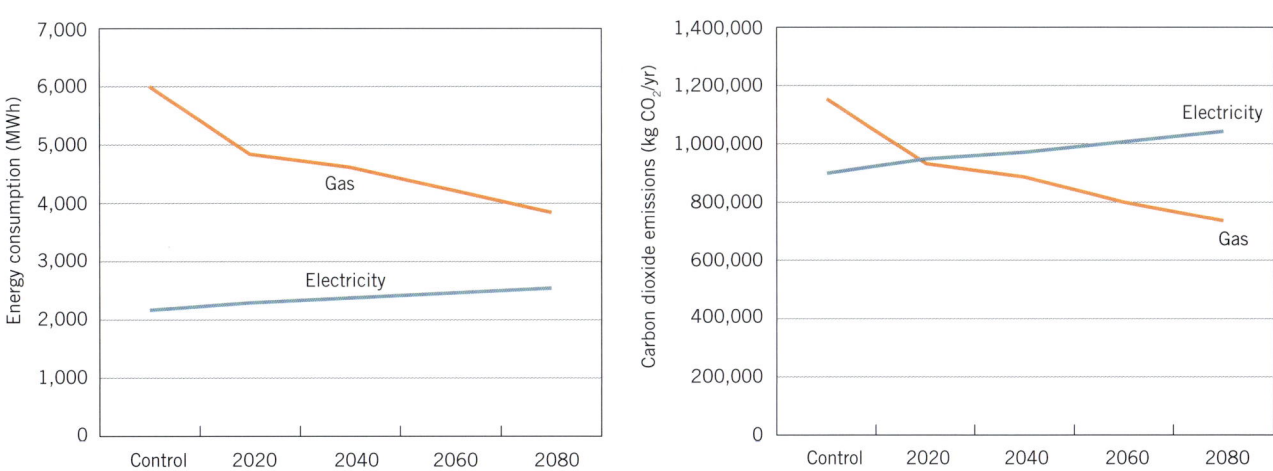

5.14 These graphs, produced by Hoare Lea for the University of Greenwich project, show how the consumption of electricty and gas may change over the coming century, and that emissions from the use of electricity will surpass those from gas as early as the 2020s.

University of Greenwich project by Hoare Lea, showing the change in energy consumption for cooling and heating and associated carbon dioxide emissions. The first chart shows that demand for gas for heating reduces, while electricity consumption for cooling increases, although gas remains the predominant load even until 2080. But when this is expressed in terms of carbon dioxide emissions, as the second chart shows, electricity is responsible for the greater proportion of CO_2 emissions as early as 2020 for this building.

In response to the long-term reduction in heating demand, and the resulting change in the balance between space heating and hot water, several teams proposed a modular approach to heating installations, so that oversized plant would not be running inefficiently as demand reduced.

The teams' forecasts for changes in cooling demand were dramatic – at 100 City Road, for example, demand for cooling is expected to be 49% higher by 2050, and 148% higher by the 2080s (albeit under the 90th-percentile High-emissions scenario).

The inadequacy of current (ie historical) weather data for the task of sizing systems is illustrated by Hoare Lea's observation for the University of Greenwich that the cooling coils, specified on the basis of current data, are likely to be too small to cope with the peak loads of a 2020s climate. It is worth noting that the 2020s climate projections represent the averages for the 30-year period centred on the 2020s – ie from 2010 to 2040, or now. It should therefore be no surprise if plant specified using industry-standard historic data struggles to cope before it reaches the end of its normal service life.

5.15 School of Architecture, Design & Construction, University of Greenwich, designed by Heneghan Peng Architects.

To accommodate additional cooling plant at Greenwich, Hoare Lea included knock-out panels for larger risers and identified the potential to install additional plant into space currently used for archives, on the assumption these would be transferred to digital format by the time it was needed. For its Oxford University Press project, the structure in one area was strengthened to support a future mezzanine for plant space. Here, the team also suggested that the cooling load could be spread by making ice overnight when electricity demand is lower, relieving the pressure on the grid at peak times and providing resilience against interruptions in power in case it was unable to cope.

At the EBI Hub, AECOM too noted the need for additional plant spaces, suggesting that modular chillers could be added as and when required. The team also recognised that cooling systems are becoming more efficient (and smaller), and suggested that as the cost of cooling and ventilation increases, higher-efficiency systems and more intelligent controls will become more viable, with the additional cost balanced against savings in plant space.

Unfortunately, the laws of physics are working in the other direction. AECOM also made the point that as external temperatures rise, the coefficient of performance of cooling systems will drop as they have to work harder to dissipate heat to a warmer environment. They also noted an increase in humidity, which will correspondingly increase the latent cooling load (the amount of energy required to dehumidify the air in a building).

There is huge potential for the development of energy-efficient low-carbon cooling systems, an opportunity that manufacturers will be quick to exploit. But there is also a huge opportunity for the proliferation of cheap, inefficient units likely to be installed as a knee-jerk reaction to a heatwave – something else manufacturers will no doubt recognise. Once installed, the only factor limiting their use will be energy price (or shortage). Of course, there could be a similar opportunity for passive measures such as awnings and blinds, though these may not be regarded as such an effective instant fix in the face of extreme temperatures.

● ● ●

EXTERNAL SPACES

Several projects noted that shaded outside spaces, water features and water bodies could become an important source of respite from high internal temperatures. To be effective, they must allow people to take advantage of any air movement as well as offering shade, particularly under trees or planting. Simple measurements using an infrared thermometer demonstrate that not only do plants provide a straightforward shading function – shielding people below them from direct sun and, indirectly, by reducing the temperature of the shaded ground – but also that the foliage is physically cooler through transpiration, helping to cool the air passing through.

The landscaping report carried out for the Church View project by landscape architect Estell Warren identified several measures that were desirable, even though they were unlikely to have any significant cooling effect on the building itself. These included planting additional trees in surrounding areas and in the central courtyard, installing climbing plants on a south-facing facade, adopting extensive roof gardens on areas of flat roof if they could be made accessible to building occupants, and installing misting devices or sprinklers in the central courtyard for use during extreme heat events: "A cooler courtyard would provide occupants in the buildings with an alternative outdoor space for use on days when extreme heat events may make internal rooms uncomfortable, in addition to providing the general benefits of an attractive outdoor space."

On the Oxford University Press project, landscape proposals aimed to make existing, undervalued courtyards and surrounding spaces more useful and inviting with external furniture and carefully selected trees to provide shade as well as enhanced biodiversity. External drinking fountains were also proposed by Hoare Lea here and for the University of Greenwich project, where the team examined the potential to create accessible, inviting roof areas and external spaces to allow the university to continue operating during periods when rooms might be uncomfortable.

5.16 Oxford University Press offices, refurbishment and extension, designed by Berman Guedes Stretton.

The team working on the British Trimmings extra-care scheme considered alternative landscaping designs that would both ameliorate the climate impact on buildings and provide an external environment to promote activity and enhance health. They also commented: "The lack of private outdoor space would be a severe limitation on occupants' lifestyles, and potentially health, in a hotter future climate where a greater proportion of their time could reasonably be expected to be spent outdoors. The use of projecting balconies would also help by providing shade to flats below."

Clearly, these suggestions would make for a more delightful place to live now, as well as in a future climate where there is likely to be more opportunity to enjoy outdoor space. They inevitably have an impact on construction costs, though some or all of this might be recouped as potential occupiers recognise the higher quality of the environment they will be buying into.

Commenting on external spaces around Trowbridge's County Hall, WSP Built Ecology noted that a simple shading device would reduce the resultant temperature for the people under it by about 5°C, from an uncomfortable 38°C to a more tolerable 33°C. At Great Ormond Street Hospital, they borrowed a useful metric from the California Energy Code Comfort Model (2008) to estimate the period of the year for which shading would be beneficial (when dry-bulb temperatures exceed 21°C with global horizontal radiation exceeding 315.5Wh/m^2).

This was one of the few instances where teams suggested specific metrics or standards from warmer places that could be applicable to a future UK climate. This perhaps reflects the international pedigree of key team members – Australians working between WSP's offices in London and New York – and may be one of the advantages of the "global village". There must be many potentially transferable metrics and standards that should be investigated in order to streamline the UK's response to climate change, and it is surprising that the appropriate government departments do not seem to be actively exploring these.

● ● ●
PUTTING IT ALL TOGETHER

It is clear that designing buildings to minimise the need for cooling in a warming climate is a juggling act. This is not just because of the need to balance a range of related impacts and strategies in order to ameliorate or avoid them, but also because, as temperatures rise, everything matters in minimising energy consumption – even relatively small factors.

It is also likely that given the moving target of rising temperatures, adaptation strategies will involve sequential but interdependent interventions that can be implemented when the evidence to justify the cost of them becomes clearer. The most effective order of implementation is, however, not always obvious.

The adaptation strategy for the Church View project was inherently sequential, to suit the building's multi-tenancy serviced-office use, and with varied packages of interventions to suit the characteristics of each type of space. The team tested different sequences of interventions to find cost-effective "easy wins" and to establish a logical order, with each intervention building on previous ones.

They provided comprehensive information for eight sample rooms (examples shown overleaf). For each they included a location plan, a three-dimensional depiction of the proposed interventions, information on modelling assumptions, a brief description of the set of potential interventions, and a bar chart showing the effect of each individual intervention on overheating.

Meeting room
1st floor
Floor area: 55m²
Ceiling height: 3.5m
South/South West facing

5.17 Bauman Lyons produced detailed outputs for each of eight sample rooms, shown on the building plan above.

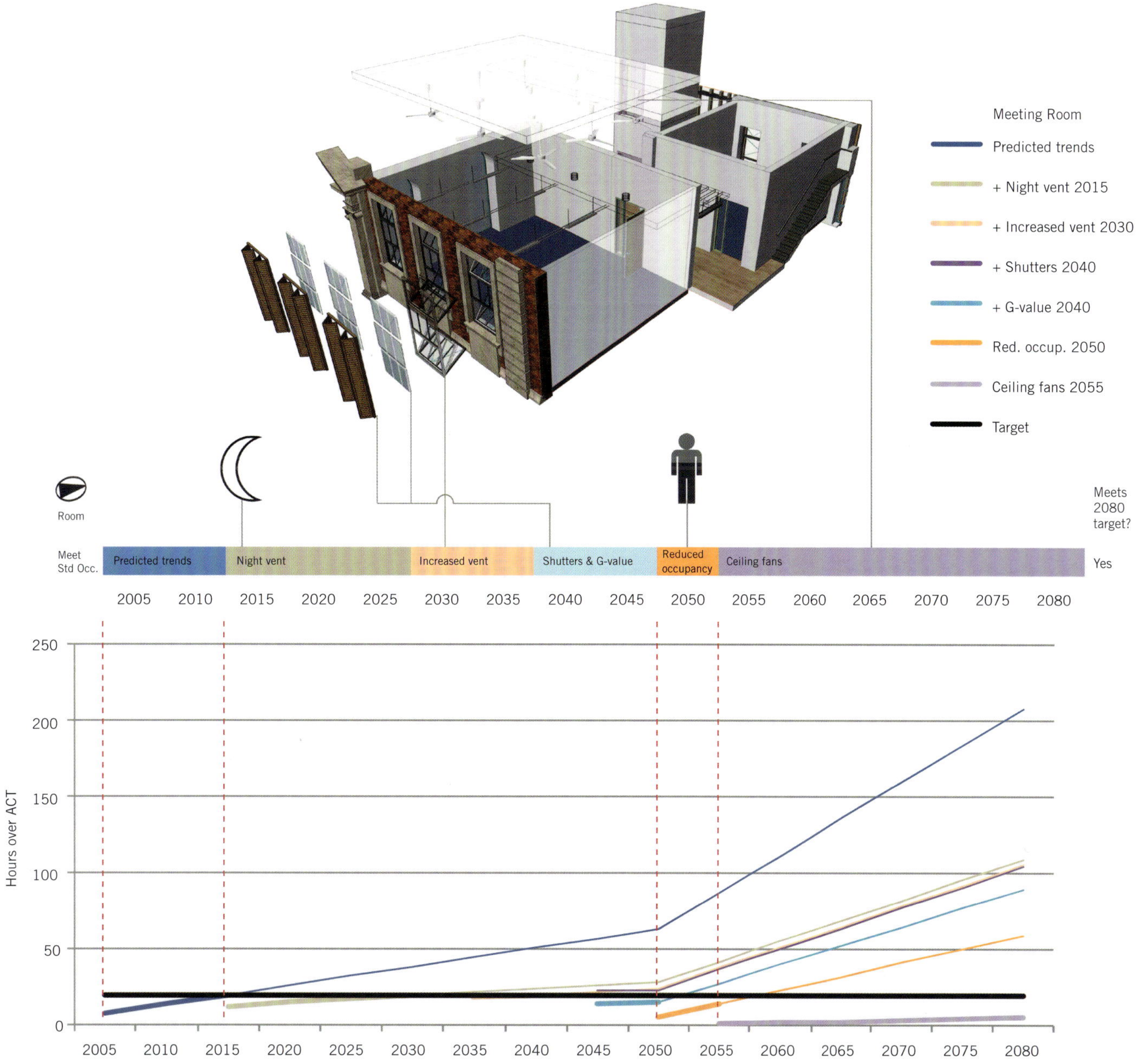

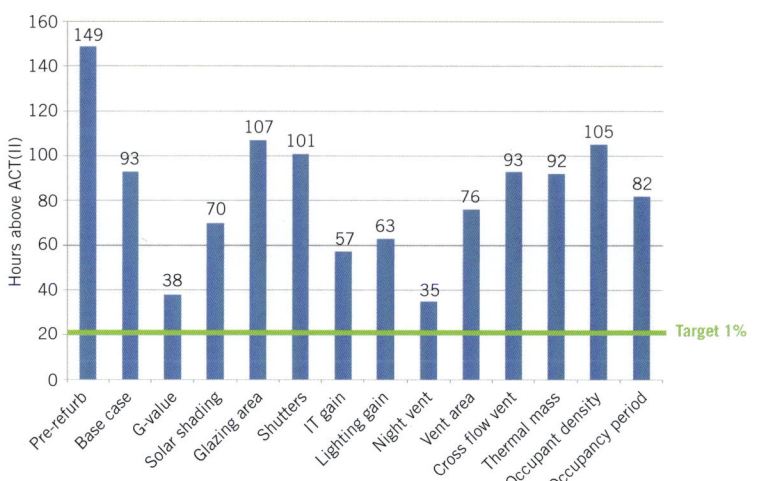

Meeting Room
Predicted trends
+ Night vent 2015
+ Increased vent 2030
+ Shutters 2040
+ G-value 2040
Red. occup. 2050
Ceiling fans 2055
Target
Room
Meets 2080 target?
Meet Std Occ.
Predicted trends
Night vent
Increased vent
Shutters & G-value
Reduced occupancy
Ceiling fans
Yes
2005 2010 2015 2020 2025 2030 2035 2040 2045 2050 2055 2060 2065 2070 2075 2080
Hours over ACT
250
200
150
100
50
0
2005 2010 2015 2020 2025 2030 2035 2040 2045 2050 2055 2060 2065 2070 2075 2080

Hours above ACT(II)
160
140
120
100
80
60
40
20
0
149
93
38
70
107
101
57
63
35
76
93
92
105
82
Target 1%
Pre-refurb
Base case
G-value
Solar shading
Glazing area
Shutters
IT gain
Lighting gain
Night vent
Vent area
Cross flow vent
Thermal mass
Occupant density
Occupancy period

5.18 Bauman Lyons created an exploded diagram of each sample room to show the proposed interventions (top). The horizontal bar chart below shows when each measure would need to be implemented. The line graph shows the effectiveness of each measure as temperatures rise over time. The thick lines indicate the length of time over which a measure maintains the temperature below the Adaptive Comfort Threshold (ACT) for the target number of hours, while the thinner lines indicate its ongoing effectiveness if the next measure in the proposed sequence were not implemented. A further vertical bar chart was included (left) to show the comparative effectiveness of each measure if applied individually in the 2050s.

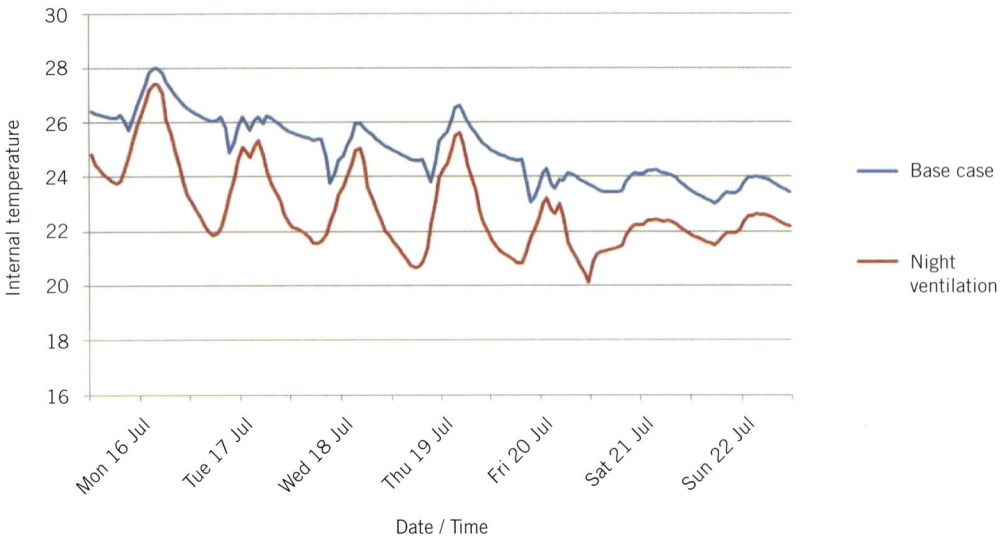

5.19 An example of additional graphical information provided for the Church View project sample rooms – in this case, the effect of night ventilation on internal temperature.

They also included a wealth of clearly tabulated information on modeling, constructional and cost assumptions. Each example included a line graph, cleverly depicting how internal temperatures were likely to change over the century using thick and thin lines to demonstrate the effects of a series of sequential interventions.

A further graph was included for each room illustrating a particularly relevant aspect of each intervention.

The proposals were summarised by combining the time-line bar charts from each sample room in a single chart as a neat way of showing the range of interventions and how and when they might be applied for these types of spaces (see page 162).

Given the need for integration, the earlier in a project's development that future summertime comfort is considered, the greater the opportunity to mould the fundamentals of the design to take advantage of available passive techniques. This is perhaps best illustrated by the Extra Care 4 Exeter project, where the adaptation study was well timed, coming just as architect Gale & Snowden inherited an initial concept design for the building, and soon after they had completed a similar project nearby. They recycled the dynamic simulation model used for the previous project and ran it with the future weather files to get a rapid understanding of the likely impacts. This gave them a more detailed insight into the likely impact of future climate on a building's performance than interpreting the raw UKCP09 data would have done at this early stage.

As a result of their analysis, they decided that one of the simplest and most effective passive approaches would be to design out single-sided ventilation completely and allow cross-flow ventilation in all of the flats. They radically revisited the scheme that had achieved outline planning consent – a typical layout with flats on either side of a central corridor – to create a curved building with dual-aspect flats arranged in three residential "clusters" around individual staircases serving just a few apartments on each level. In this, they also clearly benefited from the open-mindedness of the client and confidence that the local planning authorities would accept the change within a reasonable time-frame. The redesign not only allowed them to orient some of the individual dwellings so that bedrooms were on the relatively cooler northern facades with the living rooms facing south, but to break down a potentially institutional care home to make it a more pleasant place to live – albeit one that is, perhaps, less convenient for staff. Even though the design change was radical, because it occurred right at the start of the detailed planning design stage it did not add significant capital cost to the project.

6 CONSTRUCTION

When considering the impact of climate change on building construction, the project teams found it was much harder to obtain useful information on which to base their decisions and responses than in the case of thermal comfort.

While UKCP09 includes projections for all of the principal climate parameters that affect environmental design, there are gaps for factors that affect construction, particularly for wind and soil moisture. Not only is the required basic climatic data not available to the same level of completeness and robustness, but what information there is has not been "translated" into a form that can be used directly with familiar design standards and tools.

The 2010 report, Design for Future Climate: Opportunities for adaptation in the built environment, divided the impacts on the construction of buildings into three principal sections: structural stability, below and above ground; weatherproofing, detailing and materials; and work on site.

The report included a matrix, summarising issues to consider and the relationships between these and aspects of the changing climate. A number of the teams used this as a framework around which to structure their thinking, and it is included here for ease of reference (see figure 6.1 overleaf).

Construction

Designing for construction

This table summarises some interrelationships between anticipated changes in climate and opportunities for design, and indicates the timescales to consider when developing design strategies.[†]

Key

Climate trend
- ☀ Hotter, drier summers
- ☁ Warmer, wetter winters
- △ More extreme events

Climate information
- P Primary issue
- S Secondary issue

Time[*]
- ◔ Short – 10 years
- ◑ Medium – 25 years
- ● Long – 50 years

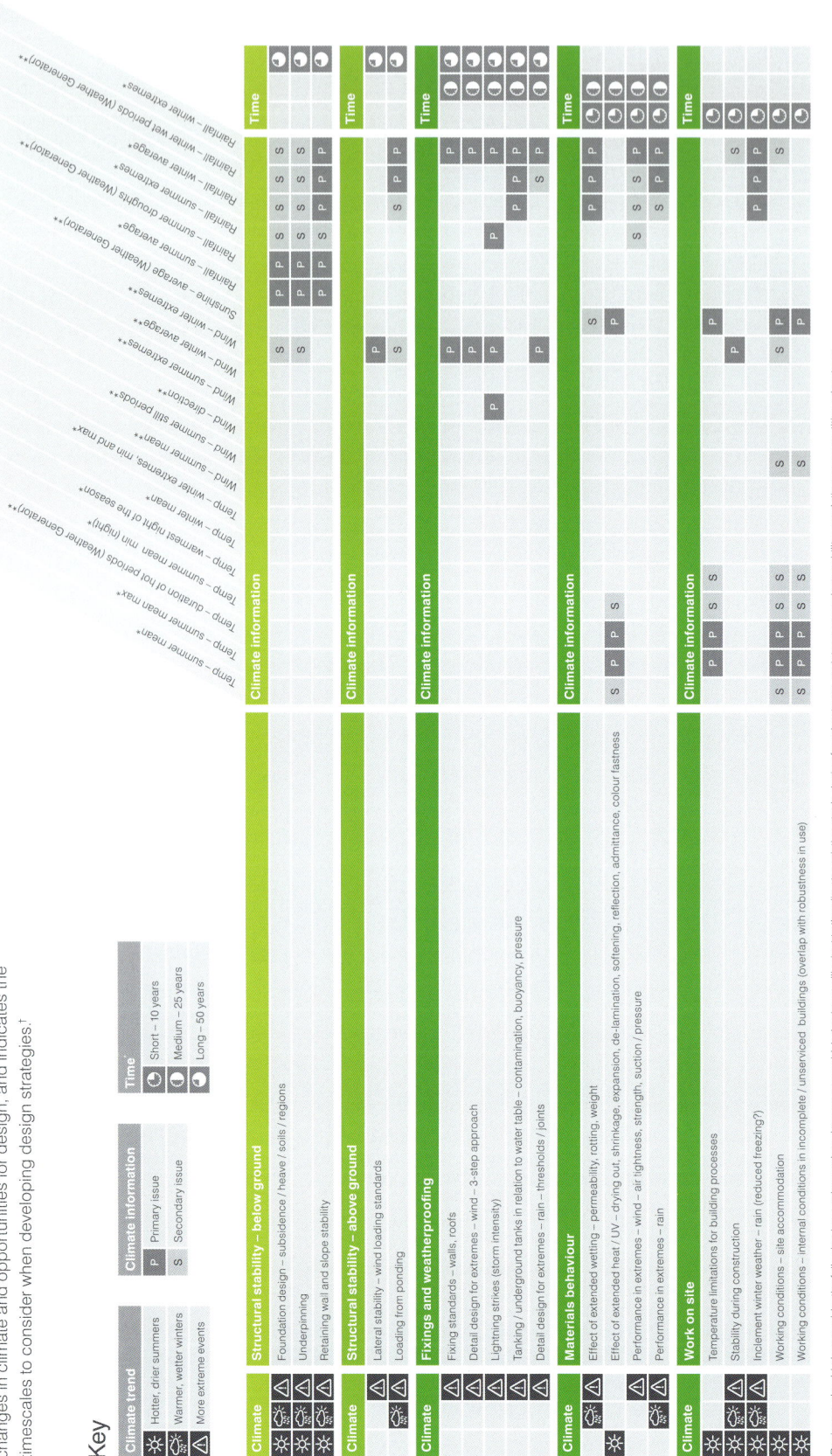

Structural stability – below ground
- Foundation design – subsidence / heave / soils / regions
- Underpinning
- Retaining wall and slope stability

Structural stability – above ground
- Lateral stability – wind loading standards
- Loading from ponding

Fixings and weatherproofing
- Fixing standards – walls, roofs
- Detail design for extremes – wind – 3-step approach
- Lightning strikes (storm intensity)
- Tanking / underground tanks in relation to water table – contamination, buoyancy, pressure
- Detail design for extremes – rain – thresholds / joints

Materials behaviour
- Effect of extended wetting – permeability, rotting, weight
- Effect of extended heat / UV – drying out, shrinkage, expansion, de-lamination, softening, reflection, admittance, colour fastness
- Performance in extremes – wind – air tightness, strength, suction / pressure
- Performance in extremes – rain

Work on site
- Temperature limitations for building processes
- Stability during construction
- Inclement winter weather – rain (reduced freezing?)
- Working conditions – site accommodation
- Working conditions – internal conditions in incomplete / unserviced buildings (overlap with robustness in use)

[†] Designers should also consider the following issues: low carbon, low energy world; behaviours will adapt to the climate; existing stock; design for robustness, maintenance and reparability; regulation vs competitive advantage; delight; regional variation.   [*] 10 years until replacement or upgrade   [*] Full probabilistic information is available from UKCP09   [**] Information is not available or only by using the UKCP09 Weather Generator

6.1 This table, taken from the original TSB report, summarises the interrelationships between aspects of climate change relating to construction and opportunities for designers, and gives an indication of the necessary timescale for strategies.

They highlighted the following potential issues in each category:

Structural stability
- impact of changing rainfall patterns on foundation design in shrinkable soils
- impact of changes in wind strength on structural design

Weatherproofing, detailing and materials
- impact of changing wind and rain patterns, and thus exposure ratings, on building detailing
- impact of changing wind and rain patterns on the behaviour of materials
- impact of changes in temperature and UV radiation on the behaviour of materials

Work on site
- impact of high temperatures and extreme rain and wind events on site processes
- impact of high temperatures on accommodation and working conditions for site staff.

Within other sectors, supply-chain vulnerability has also been identified as a potential issue. This would equally apply to the construction industry, which is just as susceptible to interruptions in the flow of materials and to difficulties concerning staff travelling to sites.

Broadly speaking, responsibility for dealing with these issues falls to different members of the professional construction team, albeit with some crossover.

● ● ●
STRUCTURAL STABILITY

This is the domain of structural engineers. They base their design proposals on British Standards or Eurocodes, and it is these against which liability for failures is judged. At present, the codes do not allow for climate change, and although teams commented on how projected changes might impact on design, they reported great difficulty in persuading clients to consider "going beyond the codes", particularly if this involved additional cost.

Foundation design

Although UKCP09 does not include specific projections for soil moisture (unlike the previous projections from 2002), the pattern of lower rainfall in summer and higher in winter is likely to increase movement of shrinkable soils. Lightly loaded domestic-scale properties with shallow foundations in these types of soils are most at risk, as larger buildings tend to be founded deeper and are less affected by changes in moisture levels near the surface. Though most of the Design for Future Climate projects concerned large new buildings, some teams did raise the issue of foundation design.

Older properties that pre-date deeper foundation design standards, triggered by the effect on existing properties of exceptionally dry summers in the latter part of the 20th century, are likely to be more vulnerable, although the most susceptible may already have been underpinned.

The inherent vulnerability of the existing domestic stock was exploited by the Trowbridge County Hall team when considering whether the conventional strip foundations of the original 1930s building might be affected by soil movements. Trial pits revealed that the existing foundations were some 2m deep, well below the depth at which most domestic properties are founded. The team reasoned that any future problems from soil movement would manifest themselves in nearby properties first – acting as an early warning system for County Hall.

6.2 Trowbridge County Hall, Wiltshire, where WSP Built Ecology developed a climate adaptation strategy for Wiltshire council.

Foundations need to be designed for the lifetime of a building, as remedial underpinning work will inevitably be significantly more expensive than a more robust foundation design at construction stage. This was the rationale followed by the British Trimmings team. Their original foundation design had been split to take account of variations across the site: part was to be founded on conventional strip footings, part on piles. As the piling rig would already be on site, the client was willing to pay a relatively small additional cost for the peace of mind of founding the entire building on piles, which would not be susceptible to soil-moisture variation.

Stability

A number of teams considered the safety factors on which existing structural design codes are based. UK structural design is governed by Eurocodes, which establish principles and requirements for both safety and serviceability. To be safe, a building must not collapse when subjected to its peak design load, a criterion referred to as the "ultimate limit state". To be serviceable, its constituent elements should not deflect by more than amounts laid down in the codes under everyday loadings, referred to as the "serviceability limit state". The serviceability limits are set to ensure that occupants are not unnerved and that architectural finishes are not damaged by excessive movements or deflections.

AECOM produced a useful summary of the design issues facing structural engineers for its EBI Hub project. This points out that the codes incorporate safety factors that take into account the likelihood that the building will experience its full design loading at some point in its life (typically around 50 years). In the Eurocodes, agreed safety factors are applied when considering live loads such as wind, temperature, rain and snow, acting both in isolation and in combination. This gives the design a good degree of robustness.

There are no safety factors for serviceability limits, which means that any increases in live loads due to climate change are likely to lead to increased movement. While this would not necessarily affect a building's structural adequacy, it could have knock-on effects for other components.

This is a fine distinction, but one that architects should note. While the ultimate failure of a building could clearly be catastrophic, excessive movements could do considerable, and costly, damage to elements such as building finishes and facade components, which are the responsibility of the architect and component designer/supplier.

Safety factors are based on the level of predictability of the materials being used, the predictability of environment in which they are being used, and a degree of redundancy in building structures. Any increase in peak live loads still reduces the safety factors on which the ultimate limit state design is based – safety factors which are, after all, there for a reason. Therefore, the peak wind speeds on which the codes are based should be reviewed, or, alternatively, consideration given to increasing the safety factors to recognise increased uncertainty.

Wind loading

Teams commented on the lack of available data in this area. Wind data was not initially included in the UKCP09 probabilistic projections, because there was considered to be insufficient correlation between climate models to create a statistically robust set of projections. Projections for wind have now been produced, but as a separate batch. This means that they cannot be correlated with other variables used in the Weather Generator projections, which presents difficulties for designers.

The pattern and strength of winter storms (and hence, the majority of wind damage) appears to be more closely related to variations in the position of the North Atlantic storm track than to changes in the climate. The historical record, rather than climate modelling per se, may therefore provide stronger evidence for making changes to design standards. It should be noted that since the first British Standard wind-loading code of practice was published in 1944, prescribed design wind loads have more than doubled for some building types and more than tripled in some parts of the UK to reflect wind speeds observed during extreme events.

That said, the engineers at AECOM, in their work on the EBI Hub, pointed out that wind loads derived from the current Eurocode 1 part 4 (BS EN 1991–1–4:2005) are felt to be conservative, at least for low- and medium-rise structures. They also pointed out that, as far as they were aware, wind damage to UK buildings is generally limited to roof and cladding fixings rather than to the structural frame – again highlighting the separation between structural matters (their concern) and components and finishes (the architect's concern).

In the case of the EBI Hub, resistance to lateral loading (ie wind) was governed not by the requirements of BS EN 1991–1–4:2005 but by the building's weight. The team concluded that, unless there was a significant increase in wind speeds, the building's structure would not be adversely affected.

For its Exeter site, Gale & Snowden highlighted slight differences between the wind-speed maps in the older, now superseded, BS 6399 part 2 on wind loads and the current Eurocode. They decided to err on the side of caution by upgrading wind loadings by 10%, adopting a recommendation from an NHBC Foundation report.[14] This was based on a proposal from the Association of British Insurers[15] that design codes for domestic buildings in the south-east of England might need to be upgraded by at least 10%, and also on the recommendations of BRE Digest 499 on the implications of climate change for the design of roofs.

Triangle Architects, working on the British Trimmings extra-care scheme, highlighted a discrepancy between the original and retrofitted elements of a design. Whereas shading devices fitted on a new building will be routinely checked for inherent strength and the integrity of fixings as part of the structure, this would not necessarily be the case for similar devices retrofitted at a later date. The team stressed the importance of providing a clear strategy and calculated provision for fixings if it is anticipated that these will need to be fitted to buildings in future.

Snow loading

Although specific projections for snowfall are not included within UKCP09, the AECOM team noted that there was a low probability of increased snowfall for their Cambridgeshire site. In this case, access was the dominant load on which the structural design of the roof was based. Recent examples of extreme winter conditions should, however, be a reminder of the need to always take account of the inherent variability of weather.

Temperature

AECOM pointed out that buildings are currently designed for the temperature ranges specified in the codes. Increases in temperature could potentially affect structures, through expansion and contraction, but no attempt was made to quantify the possible impact on other building components (generally not specified by structural engineers). Product standards may need to be reviewed to take account of changes in the temperature band within which components will have to perform. Specifiers would be wise to add tolerance to higher temperatures to their list of considerations when selecting products.

On structural materials, AECOM made the following observations.

- Steel actually gains strength with rising temperatures. A warming climate is also likely to reduce (marginally) the brittleness that can occur with exposed steel at low temperatures.
- Concrete pours are likely, on average, to be interrupted less frequently by freezing conditions (due to warmer winters), although extreme conditions may still occur. During warmer summers, when temperatures rise above 23°C, the risk of cracking can be reduced by altering the cement-to-water ratio or by including additives in the mix. Above around 35°C, however, safe curing of concrete requires special measures.

One of the advantages of Eurocodes appears to be that the same methodology is used across a wide range of existing climates, albeit with local climate parameters. One would imagine that it is relatively simple to adapt British design and construction practice to reflect tried-and-tested methods in regions which, for example, currently experience higher temperatures. Similarly, UK engineering companies have extensive experience of building in warmer climates, such as the Gulf states. They should draw on this experience when considering changing conditions at home.

The factors of safety built into structural design appear to offer some breathing space while changes to codes and standards are considered. As the UK design community relies so directly on such codes and standards, the impacts of climate change are unlikely to be considered until those changes are put in place. Indeed, the community seems to be putting the onus firmly on those who control the standards, rather than developing ad hoc strategies on a project-by-project basis. It is therefore important that the standard-setting bodies consider how, if at all, potential impacts should be addressed when compiling future editions of the codes. That said, it is interesting to speculate as to where the courts will allocate blame in situations where a failure has occurred even though the designers complied with the codes in use at the time.

● ● ●
WEATHERPROOFING, DETAILING AND MATERIALS

Among the TSB projects, the most systematic analysis of the impact of climate change on building fabric was carried out by Arup for the Engineering Graduate School at the University of Sheffield, as part of a comprehensive facade study.

6.3 Engineering Graduate School, University of Sheffield, designed by Bond Bryan Architects.

The full report included solar studies using Ecotect software to investigate appropriate G-values for glazing in different locations on the building, and an evaluation of alternative facade systems that considered a range of issues including service life, maintenance, weathering, embodied carbon and lifetime costs. The study was extended to carry out a qualitative assessment of each option against the impacts noted in the original Design for Future Climate report, namely:

- hotter, drier summers
- warmer, wetter winters
- more intense rainfall events
- greater temperature extremes
- increased wind speed
- prolonged UV exposure.

A system of traffic-light colour coding was used to flag up levels of vulnerability and resilience (see figure 6.4 overleaf). Although the evaluations were not comprehensive, a number of widely applicable issues were noted:

- a need to accommodate greater levels of thermal expansion, particularly where metal facade materials are bonded to backing boards
- the role of thermally massive materials in reducing peak temperatures on the surface of a facade
- greater saturation of facade materials, potentially leading to increased organic growth and, in extreme cases, to water being driven through to cavities or insulation layers
- increased rain penetration of rain-screen systems, putting greater strain on joints and backing materials, which will require adequate drainage and barriers
- higher levels of facade erosion as a result of driving rain
- increased likelihood of warping of timber-based boards
- a need for additional fastenings for cladding systems
- more rapid fading of colours as a result of increased UV radiation.

The review found that although all systems would suffer to some degree, the effects could be satisfactorily managed with "appropriate consideration at design and proper specification".

At Great Ormond Street Hospital, WSP Built Ecology proposed that the exposure rating for the facade design should be increased from 1 to 2 in order to allow for climate change. The team made a similar recommendation for the Trowbridge County Hall project (from 3 to 4 in this case), and it was decided not to inject thermal insulation into the cavities of the existing building due to an unacceptable risk of moisture penetration through the existing stonework.

Gale & Snowden opted to use a "very severe" exposure rating for construction detailing on the Extra Care 4 Exeter project, taking a similarly cautious approach to this element of the

		HOTTER, DRIER SUMMERS	WARMER, WETTER WINTERS
CURRENT OPTION	Stone rainscreen fixed onto Metsec backing wall	Increased temperatures will cause more thermal expansion which could result in increased risk of cracking if there are not sufficient movement joints.	As the stone becomes more saturated there is an increased likelihood of organic growth. However, the impact of this would be primarily aesthetic.
ALTERNATIVE 1	Through coloured insulated render system	Increased temperatures will cause more thermal expansion. Renders have high expansion capability so this should not pose a problem, but the increased temperature range should be considered at the design to ensure an appropriate render is chosen.	As the render becomes more saturated there is an increased likelihood of organic growth, as the additives only provide a level of protection. However, the impact of this would be primarily aesthetic.
ALTERNATIVE 2	Natural copper cladding on 20mm marine ply	Increased temperatures will cause more thermal expansion. The copper will expand more due to the temperature than the timber, therefore the bond between these elements will need to be carefully specified to allow for these stresses.	Increased humidity due to warm, wet air will make the plywood more susceptible to degradation.
ALTERNATIVE 3	Trespa panels (ventilated facade system)	No effect. Trespa boards have high dimensional stability at elevated temperatures.	No effect. Trespa boards have good resistance to moisture.
ALTERNATIVE 4	Load-bearing brick	Thermal mass of both the brick and block elements of the wall may be able to alleviate extreme temperatures. Increased temperatures will cause more thermal expansion which could result in increased risk of cracking.	As the bricks become more saturated there is an increased likelihood of organic growth. However, the impact of this would be primarily aesthetic.
ALTERNATIVE 5	Natural red stone cladding (solid wall construction with cavity and inner face blockwork)	Increased temperatures will cause more thermal expansion which could result in increased risk of cracking if there are not sufficient movement joints. Thermal mass of block work wall may be able to alleviate extreme temperatures.	As the stone becomes more saturated there is an increased likelihood of organic growth. However, the impact of this would be primarily aesthetic.
ALTERNATIVE 6	Lignacite concrete facing masonry block	Thermal mass of block work wall may be able to alleviate extreme temperatures. Increased temperatures will have no effect on the durability of the concrete facing blocks.	As the concrete becomes more saturated there is an increased likelihood of organic growth. However, the impact of this would be primarily aesthetic.

6.4 Arup investigated the resilience of a range of facade systems for the University of Sheffield project, using a traffic-light system to compare the performance of each against the climate impacts noted in the original Design for Future Climate report. Key to text colours: orange = minor vulnerability, black = no effect, green = resilience

SHORTER PERIODS OF MORE INTENSE RAIN	GREATER TEMPERATURE EXTREMES	INCREASED WIND SPEED	PROLONGED UV EXPOSURE
More intense rainfall will cause more moisture to be forced through the rain screen. This will require adequate drainage and barriers. Additionally if a soft stone is used, heavier rainfall will increase the rate of erosion, requiring additional thickness.	More instances of temperatures falling below zero would increase the likelihood of failure due to freeze/thaw action. This would need to be covered in design detailing and specification.	If a soft stone is used, increased wind speeds will increase the rate of erosion; however this will still be minimal.	No effect.
More intense rainfall will cause more moisture to be forced through the render to the substrate below. This will require adequate drainage and barriers.	More instances of temperatures falling below zero which would increase the likelihood of failure due to freeze/thaw action. This would need to be covered in the specification.	Increased wind loads may require the use of additional fastenings or, where the system is adhesively fixed, the addition of mechanical fixings.	Coloured renders may fade more rapidly; however they are fairly stable.
More intense rainfall will cause more moisture to be forced through the rain screen. This will require adequate drainage and barriers.	The panels are unlikely to suffer from freeze/thaw action; however they may be liable to warping.	Increased wind loads may require the use of additional fastenings, this would need to be considered in the design process.	No effect.
More intense rainfall will cause more moisture to be forced through the rain screen. This will require adequate drainage and barriers.	As the panels are very thin they may be susceptible to warp when put under the stresses caused by greater temperature extremes.	Increased wind loads may require the use of additional fastenings; this would need to be considered in the design process.	No effect. Trespa claim their panels are resistant to UV degradation, and have tested them based on a Florida weather cycle. Caution however: UV exposure can be a problem.
More intense rain will increase the risk of the mortar being eroded. Either a mortar with more cement would need to be used or the mortar would have to be re-pointed more often. Driving rain and a more humid climate will increase the risk of water penetration.	More instances of temperatures falling below zero which would increase the likelihood of bricks failing from freeze/thaw action. This would require the specification of more frost-resistant bricks.	Increased wind speeds will increase the risk of the mortar being eroded. Either a mortar with more cement would need to be used or the mortar would have to be re-pointed more often. Driving rain and a more humid climate will increase the risk of water penetration.	No effect.
More intense rainfall will cause more moisture to be forced through the rainscreen. This will require adequate drainage and barriers. Additionally if a soft stone is used, heavier rainfall will increase the rate of erosion.	More instances of temperatures falling below zero which would increase the likelihood of failure due to freeze/thaw action. This would need to be covered in design and specification.	If a soft stone is used, increased wind speeds will increase the rate of erosion; however this will still be minimal.	No effect.
More intense rain will increase the risk of the mortar being eroded. Either a mortar with more cement would need to be used or the mortar would have to be re-pointed more often.	No effect. It is very unlikely that the concrete will suffer due to freeze/thaw action.	Increased wind speeds will increase the risk of the mortar being eroded. Either a mortar with more cement would need to be used or the mortar would have to be re-pointed more often.	No effect.

building as they had to the structural design. As a result, the team selected a StoTherm external insulation-and-render system that is frequently used in countries with significantly more extreme weather conditions than the UK. In consultation with the supplier, they modified their standard design practice to include:

- using a silicon-resin-based render
- stooled ends to sills
- increased overhangs and vertical legs of flashings.

They also established that the fixing system for the render gave them an extremely high factor of safety (greater than 21) based on a worst-case wind-suction scenario – all at practically no additional cost.

Changing patterns of rainfall could have severe repercussions on the existing stock, when combined with internal insulation fitted as part of a thermal upgrade (as, for example, under the government's Green Deal scheme). This was not covered by any of the TSB-funded projects, but it was addressed as part of a comprehensive thermal upgrade of student rooms at Trinity College Cambridge carried out by 5th Studio and Max Fordham with input from English Heritage, Archimetrics and Natural Building Technologies. Using WUFI software to model the movement of water in liquid and vapour through the building's existing solid masonry walls,[16] they concluded that moisture penetration due to increased future rainfall was the limiting factor in determining the thickness of internal insulation that could be applied without potential long-term damage to timbers buried in the wall.

Cambridge is an area of relatively low rainfall, and the walls in question were comparatively thick – at least 450mm. For a more typical 225mm domestic solid wall in a wetter part of the country, the "safe" level of internal insulation will be correspondingly reduced. It is vital that this issue is investigated properly and that reliable simulation software is developed and tested against reality, so that robust decisions can be made. Without this understanding, there is a danger that thermal upgrading will result in unintended negative consequences and, at worst, wholesale damage to historic fabric.

WORK ON SITE

As the funded projects were all live and programmed for construction in the current climate, observations of the impact of climate change on site operations tended to be general conjectures of what might happen in future, rather than specific issues that had been raised during the research.

Most changes to site working practices and conditions would clearly be the responsibility of building contractors, a part of the industry that was not generally involved in the Design for Future Climate projects. It should, however, be noted that contractors are already starting to consider summer climate-related issues on building sites. A recent "toolbox talk" initiative by Simons Group was initially prompted by the prospect of water shortages in the summer of 2012 and the difficulty of controlling site dust under those circumstances. But the talk also covered the dangers of overheating and heat stress for construction staff while buildings were under construction.

Under CDM regulations (Construction Design and Management legislation dealing with building-safety issues), the design team has a responsibility to flag up threats to health and safety during site operations. They will need to consider whether there are factors inherent in the design that make it difficult to control internal environmental conditions in the half-completed building. Given that conditions similar to those anticipated in the UK already exist in other parts of Europe, we should take the initiative to make ourselves aware of how potential problems can be overcome before they arise.

Designing to manage water

This table summarises some interrelationships between anticipated changes in climate and opportunities for design, and indicates the timescales to consider when developing design strategies.[†]

Key

Climate trend
- ☀ Hotter, drier summers
- ❄ Warmer, wetter winters
- △ More extreme events

Climate information
- P Primary issue
- S Secondary issue

Time
- Short – 10 years
- Medium – 25 years
- Long – 50 years

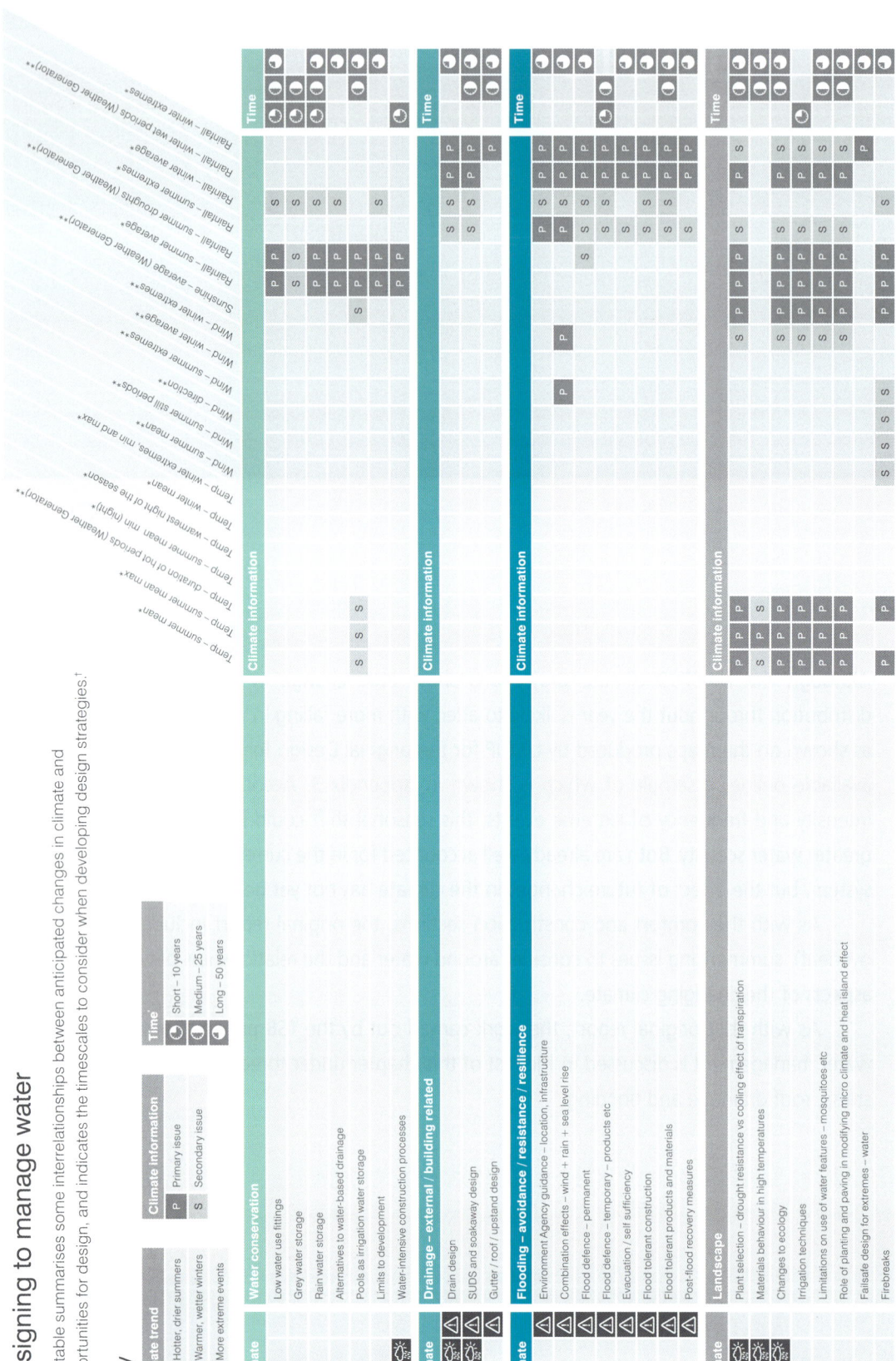

7.1 This table, taken from the original TSB report, summarises the interrelationships between aspects of climate change relating to managing water and opportunities for designers, and gives an indication of the necessary timescale for strategies.

[†]Designers should also consider the following issues: low carbon, low energy world; behaviours will adapt to the climate, existing stock, design for robustness, maintenance and reparability, regulation vs competitive advantage; delight; regional variation. ¹10 years until replacement or upgrade. *Full probabilistic information is available from UKCP09. **Information is not available or only by using the UKCP09 Weather Generator

WATER STRESS

Parts of the UK already suffer from water stress – particularly in the south and east, where there is less water available per person than there is in many Mediterranean countries (see figure 7.2 below).

Water-efficiency targets have been included in environmental standards such as BREEAM and the Code for Sustainable Homes for a number of years, and the Building Regulations have set minimum standards since 2010.

All the Design for Future Climate projects were designed to either BREEAM Excellent or Very Good, or Code Level 5 or 6, and, as a result, already included the specification of efficient appliances. Some teams also proposed alternatives to the conventional mains-water supply, involving separate potable and non-potable pipework systems.

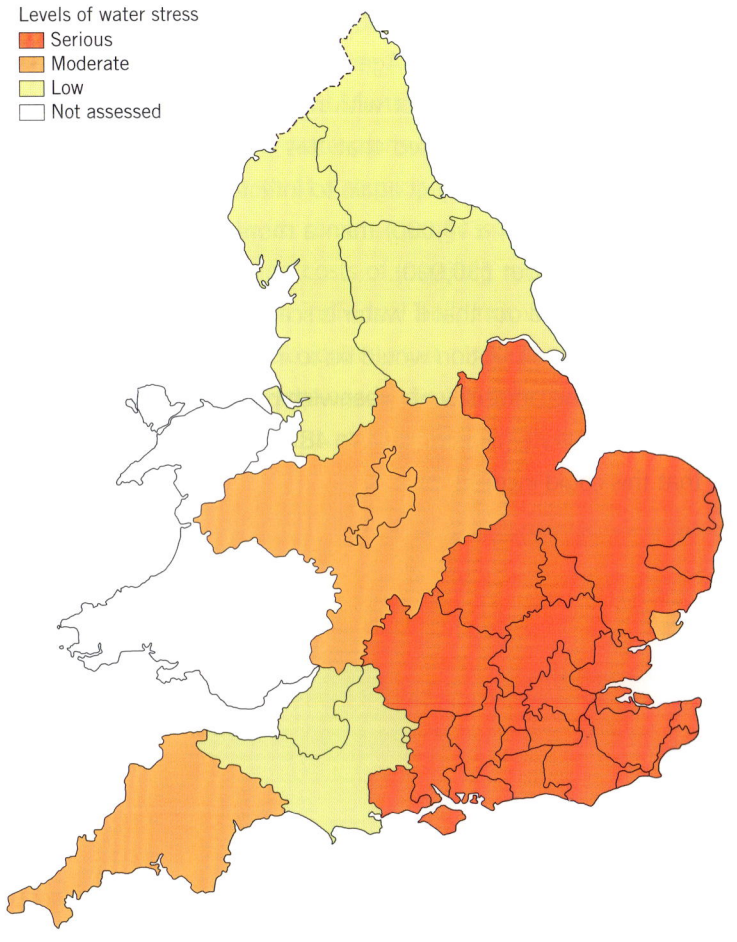

Levels of water stress
- Serious
- Moderate
- Low
- Not assessed

7.2 This map, produced by the Environment Agency, shows areas of water stress in England.
Source: Areas of Water Stress: final classification, Environment Agency, 2007, p5

● ● ● HOW BIG SHOULD A GUTTER BE?

The intensity of a rainfall event is described in terms of its expected frequency, or "return period". For example, in an analysis of 100 years of rainfall data, a 1-in-20-year event would be expected to occur five times within that data-set, and on average once every 20 years. In any given year, the probability of such an event occurring would be equal to 1 in 20, or a 5% chance. A 1-in-1,000-year storm would obviously be much more intense, but there would only be a 0.1% probability of it happening.

To work out exactly how intense an event a building should be able to withstand, designers must decide how severe the consequences of the system overflowing would be, and choose one of four risk categories. A building with conventional eaves and gutters that can overflow with no significant impact would fall into a lower category than a building with internal gutters, where an overflow could cause significant internal damage or affect some critical process. For each category, there is an associated multiplier which is applied to the anticipated life of the building to give an appropriate return period on which to base the design.

A series of contour maps provides design rainfall intensities in litres/second across the UK for return periods of 1, 5, 50 and 500 years. There is also an additional map showing maximum probable intensities for buildings which require the highest level of security. The map with a return period equal to or greater than the design return period should be chosen.

For example, a building with valley or parapet gutters would fall within Category 2, which has a multiplier of 1.5. For a building with an anticipated life of 30 years, the design return period is 45, and so the 50-year map should be used to set the appropriate design rainfall intensity.

The standard does not include allowances for climate change, and many teams noted that UKCP09 data on projected rainfall is only available at a much coarser level of detail than that needed for roof-drainage design (in mm/day rather than l/sec). There is no reliable correlation between daily averages and intense events, and so there are difficulties in making an appropriate allowance for climate change in roof-drainage design.

Gale & Snowden's rainwater-disposal design was informed by research from the University of Exeter, which used the Weather Generator to study the potential change in return periods. It found that a 1-in-100-year rainfall event could become a 1-in-50-year event as soon as 2030.

It also cited a study carried out by the Met Office for Ofwat, which analysed future rainfall return periods for 40 towns and cities across the UK using regional climate projections alongside the UKCP09 figures.[18] This publication used a different dataset to the Weather Generator but came to broadly similar conclusions, predicting that a current winter 1-in-100-year daily-rainfall event in London would become a 1-in-42-year event by the 2040s.

They therefore proposed that design return periods indicated by the British Standard methodology should be doubled in order to account for climate change.

The team working on the Harris Academy in Purley, south London, chose to use a more onerous BS EN 12056 category to give an additional margin of safety. This led to an increase in design rainfall intensity of 30%, accommodated by increasing the size of downpipes from 100mm diameter to 150mm at minimal additional cost. The cross-sectional area of a pipe, and therefore its capacity, is proportional to the square of its diameter; a 50% increase in diameter increases its cross-sectional area by 125%.

PPS 25 (Planning Policy Statement 25: Development and Flood Risk) also suggests a 30% uplift as a precautionary allowance in peak rainfall intensity for the period up to 2115. Although PPS 25 was replaced by the simplified National Planning Policy Framework in 2010, along with all Planning Policy Statements, the guidance it contains remains useful in the absence of any alternative authoritative source (see figure 7.3 overleaf).

A number of teams modified the design of gutter systems so that they would overflow clear of the building under extreme conditions, to avoid causing possible internal damage. However, it was noted that the overflowing water cannot be ignored – it has to go somewhere once it hits the ground, and so must be taken into account in surface-water-drainage schemes.

Marine flooding

PortZED, a mixed-use Code for Sustainable Homes Level 6 seafront development in Brighton and Hove, where the lower area of the site is level with the existing dock front, was the only project that needed to address marine flooding.

Architect ZEDfactory found that under most climate change scenarios, rising sea levels would flood the existing dock edge regularly within the next 50 years, with a policy of planned abandonment required by the end of this century. In response, the design team located all high-value retail and residential accommodation at the higher Kingsway street level, effectively two storeys higher than the dock. The lower two storeys contain parking, car pools and commercial space for employment, which could be sacrificed and relocated should flooding become so persistent as to make using this space untenable (see figure 7.4 opposite).

As flooding becomes increasingly frequent at dock level, it is anticipated that the whole harbour area would be artificially raised with fill, probably ending up a full storey higher by the end of the century. This strategy had already been implemented at ZEDfactory's Jubilee Wharf project in Penryn, Cornwall, another key urban waterside site designated for planned abandonment by the Environment Agency. Here, the whole development is raised 1200mm higher than the existing street in anticipation of future increases in road level.

At PortZED, the architect took care to allow for the continued operation and relocation of woodchip deliveries and storage, rainwater-harvesting tanks and biomass-fuelled boilers. The loss of non-critical areas of private parking was planned in anticipation that smaller electric vehicles and pool cars would reduce the space required, while allowing vertical services risers and energy distribution to the development above to remain functional with minimum disruption.

7.5 At the PortZED development facing Shoreham Harbour in Brighton, sacrificial car-park levels were included to accommodate flooding.

Fluvial flooding

AECOM's EBI Technical Hub site is adjacent to the River Cam. According to flood maps from the Environment Agency, only the area immediately adjacent to the river is prone to flooding, while a small part of the site falls within the 1-in-1,000-year flood envelope. The proposed location of the Hub is outside the flood zone.

The AECOM team expressed a view that the flood-risk methodology contained in Planning Policy Statement 25 is inflexible and imprecise, and does not allow the use of more detailed information available within UKCP09. PPS 25 suggests adding a percentage to peak rainfall intensity and peak river flow in order to test the sensitivity of a design to climate change (see figure 7.3 on page 137), and assumes a linear relationship between changes in rainfall and river flows. This does not correspond to UKCP09 projections for aspects such as soil moisture, groundwater baseflow and groundcover.

They also noted that while the figures in PPS 25 are intended to account for a predicted increase in 1-in-100-year events by the 2080s, they fall some way short of the changes in return periods anticipated by the Met Office report for Ofwat (see page 135).

The team therefore decided to investigate how additional information from UKCP09 could be included within flood-risk assessments, and developed an alternative approach to assess whether the Hub building would remain at a low risk throughout its lifetime.

The starting point for flood-risk modelling is the Flood Estimation Handbook (FEH), produced by the Centre for Ecology and Hydrology, which provides information such as catchment area, slope and soil type for areas across the UK, and the Revitalised Flood Hydrograph (ReFH) model, which shows how quickly rainfall will drain away across a site.

AECOM noted that the conventional approach assumes that rainfall will run off in a uniform fashion across the entire catchment. In reality this is very unlikely, even in a small catchment. Neither does it allow for the impact of changes to the physical characteristics of a site, either from climate change or other temporal effects. For example, recent studies have shown that the increase in field size from 1958 to 2008 has reduced the time to peak run-off and increased peak discharge,[19] and that changes in rainfall frequency can affect soil-moisture deficit levels, which in turn can affect run-off rates.[20] As a result, they felt that the response of their catchment area to changes in climate and physical conditions may be significantly underestimated.

AECOM's alternative approach involved updating the Flood Estimation Handbook's description of their site in advance of generating hydrographs, in order to model the impact of future changes on its response to rainfall events. They used data from UKCP09 to qualitatively inform the amendments, in order to account for ground conditions affected by previous rainfall events as well as projected urban expansion, including development in river channels.

They also modelled the impact of pairs of flooding events occurring in quick succession. Current best practice in flood modelling seeks to understand the impact of progressively more severe storms, but these are usually considered in isolation. With the predicted increase in storm events across the UK, there are likely to be occasions when one storm is immediately followed by another. When a second event hits, river flows may already be running close to or higher than their banks, soils may be completely saturated and flows may be blocked or restricted by significant amounts of debris. This means that the second storm, which might be of lower intensity, could result in significantly greater flooding than if it took place in isolation.

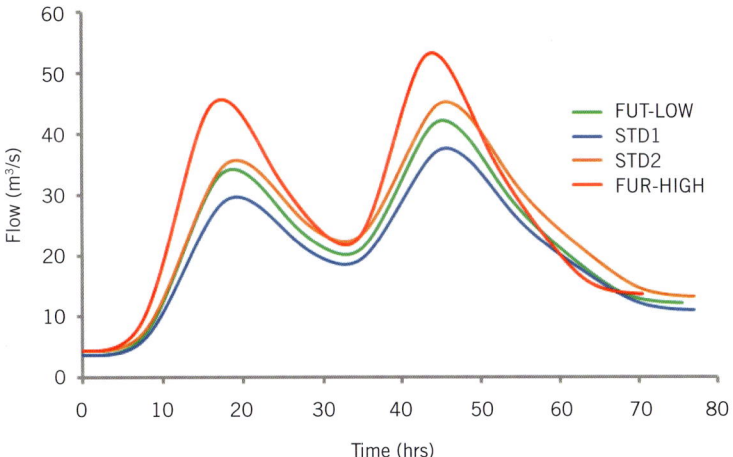

7.6 For the EBI Hub project beside the River Cam, AECOM created hydrographs showing flow rates for pairs of flooding events occurring in quick succession.

AECOM modelled an extensive set of scenarios, as follows.

1. Current 1-in-100-year rainfall event.
2. As scenario 1, but with an additional 20% for climate change (ie PPS 25 approach).
3. A future Low-emissions scenario rainfall event, using a current 1-in-125-year event as a future 1-in-100-year event (to reflect low climate-change impacts on catchment characteristics).
4. A future High-emissions scenario rainfall event, using a current 1-in-150-year event as a future 1-in-100-year event (to reflect quite substantial climate-change impacts on catchment characteristics).
5. A 1-in-50-year rainfall event followed 12 hours later by a 1-in-75-year rainfall event.
6. A 1-in-50-year rainfall event followed 12 hours later by a 1-in-75-year rainfall event, both with an additional 20% to account for climate change.
7. A 1-in-75-year storm event followed 12 hours later by a 1-in-100-year storm event using the same amended catchment descriptors as for scenario 3.
8. A 1-in-100 year storm event followed 12 hours later by a 1-in-125-year storm event using the same amended catchment descriptors as for scenario 4.

As UKCP09 contains much more detailed location-specific information (rather than simply applying a blanket 20% or 30% increase across the country), AECOM felt that incorporating this data in flood-risk assessments would produce more realistic and robust models, and enable wider catchment-management strategies.

As a result of this modelling, the team were able to show that the Technical Hub is not at risk of flooding, either now or when the impact of climate change is taken into account. But they also found that the western edge of the site appeared to be at a greater-than-1% risk of flooding in any given year – rather than 0.1%, as outlined in the EA flood maps. The back-to-back storms scenarios did not produce significantly greater flooding extents, but did demonstrate the protection provided by an upstream railway bridge during the highest flows, constricting the channel and throttling peak flows entering that section of the river.

One limiting factor on this approach was the computational burden of the more detailed modelling exercise – the team noted this was a general issue with 2D hydrodynamic models. They suggested that further work could use detailed cross-sectional surveying of the River Cam, allowing a much faster 1D model to be built. This would allow more accurate modelling of the capacity and flow in the Cam, and could be linked to the 2D model to maintain the detail of modelling produced so far.

Architect White Design produced a Future Design Code for the site of The Mill development alongside the River Ely in Cardiff. This addressed current river flooding by incorporating a river walk, a lower-lying amenity area that would be allowed to flood. This created a set of inlets integrated into the street hierarchy and a Sustainable Drainage Strategy (SuDS) to allow stormwater run-off to be controlled on a site-wide basis. The team suggested that over time the area would flood with increasing frequency and that its use might change – it could initially be used for growing food and become recreational later, before finally being given over to wetland wildlife.

7.7 The Mill, Cardiff, where White Design produced a Future Design Code, taking into account the risk of river flooding.

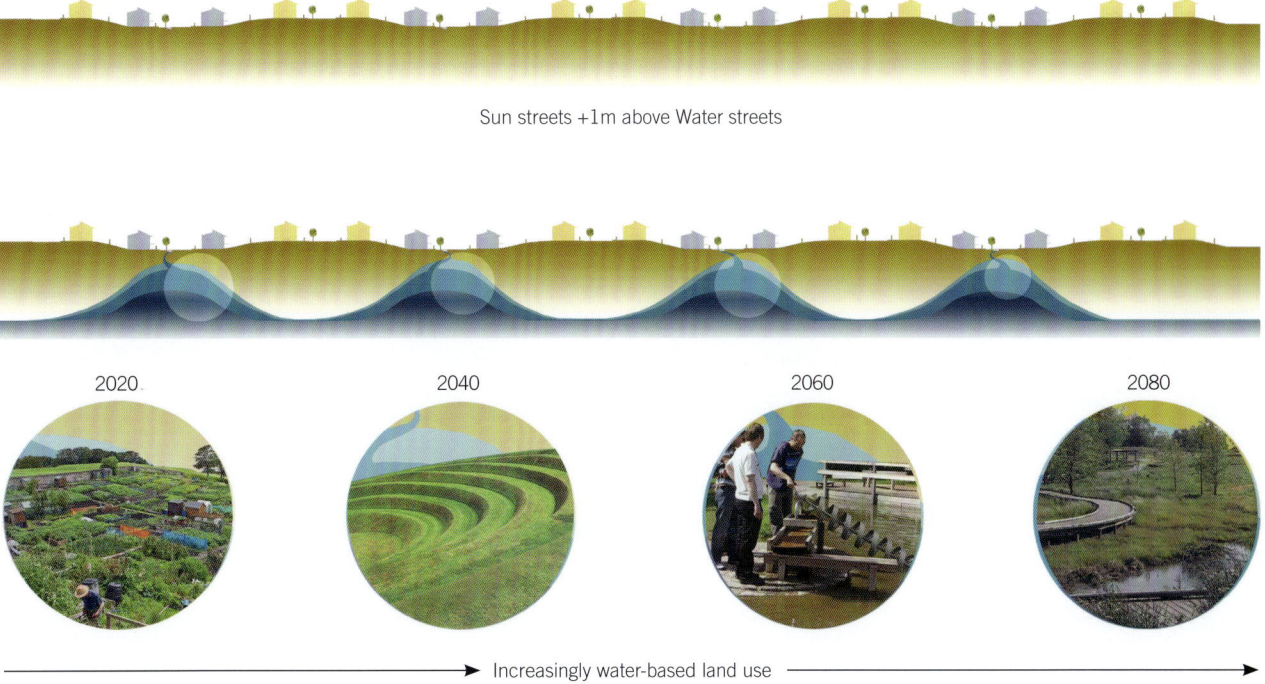

Sun streets +1m above Water streets

| 2020 | 2040 | 2060 | 2080 |

Increasingly water-based land use

7.8 White Design's land-use strategy for The Mill development alongside the River Ely evolves over time as flooding becomes more frequent.

Similarly, the team working on the new Bicester EcoTown included "eco-corridors" within their SuDS, not only to deal with intense events but as an opportunity to enhance the ecology and improve the water flow throughout the year.

7.9 Bicester EcoTown, Oxfordshire.

Surface-water flooding

Surface-water flooding occurs when the volume of rainfall exceeds the local drainage capacity. It is generally an urban phenomenon: an increasing amount of hard surfaces means that rainfall runs off rapidly rather than being absorbed by open ground, causing flooding if local sewers are not able to cope. Designers will be familiar with an expectation to adopt sustainable urban-drainage principles to reduce these risks, dealing with surface water within site boundaries as far as possible.

Surface-water flooding tends to be very localised and happens very quickly, making it difficult to provide much warning to building occupants. It is less well-understood than recurring coastal or river flood risks, and whereas there is nationally available mapping for river and sea flooding, a Flood Map for Surface Water has only recently been made available, and only to Lead Local Flood Authorities (now responsible for the management of local flood risk with the support of the Environment Agency), Local Resilience Forums, and local planning authorities.

The Flood Map for Surface Water is a useful starting point, giving a broad indication of the areas likely to be at risk of surface-water flooding. But there are strong caveats accompanying its use, because the risks are so heavily dependent on the detailed hydrology of a particular site. As with river- and sea-flooding maps and standards, no allowance is included for climate change.

BS EN 752–4:1998 (Drain and Sewer Systems Outside Buildings) uses a concept of return periods similar to that used in roof-drainage design. The points raised above by the University of Exeter's research on the effect of climate change on return periods also apply here.

The Environment Agency normally requires that surface-water drainage be based on a 1-in-100-year storm event, plus an allowance for climate change in accordance with the recommendations of PPS 25. For a design with an anticipated lifetime beyond 2085, for example, this would be an increase in peak rainfall intensity of 30%. Hoare Lea, working on the University of Greenwich project, noted that this is more onerous than the increase in rainfall intensity indicated in the climate files used (Medium-emissions COPSE files produced by Manchester University). It is unclear whether this reflects the weakness of the Weather Generator in projecting extreme events or whether the Environment Agency's precautionary allowance is excessive, and further study to investigate this would be useful.

Hoare Lea's discussions with the client highlighted their potential vulnerability to surface-water flooding, and also the fact that the surface-water-drainage system in London is not yet well understood.

Surface-water flooding was found to be the principal risk at Great Ormond Street Hospital. An upgrade of the drainage system was already planned as part of the project and, by using LIDAR (Light Detection And Ranging) topographical information to inform their understanding of surface-water flows in the area, the engineers were able further to reduce the likelihood of surface-water flooding. While the probability of flooding was assessed as low, the potential impact would be very high because the building would need to remain operational during such events. The WSP Built Ecology team therefore proposed a combination of dry- and wet-proofing approaches to reduce the vulnerability of basement areas and the surface-water design was modified so that should drainage be overwhelmed in future extreme events, water could flow away over the surface safely.

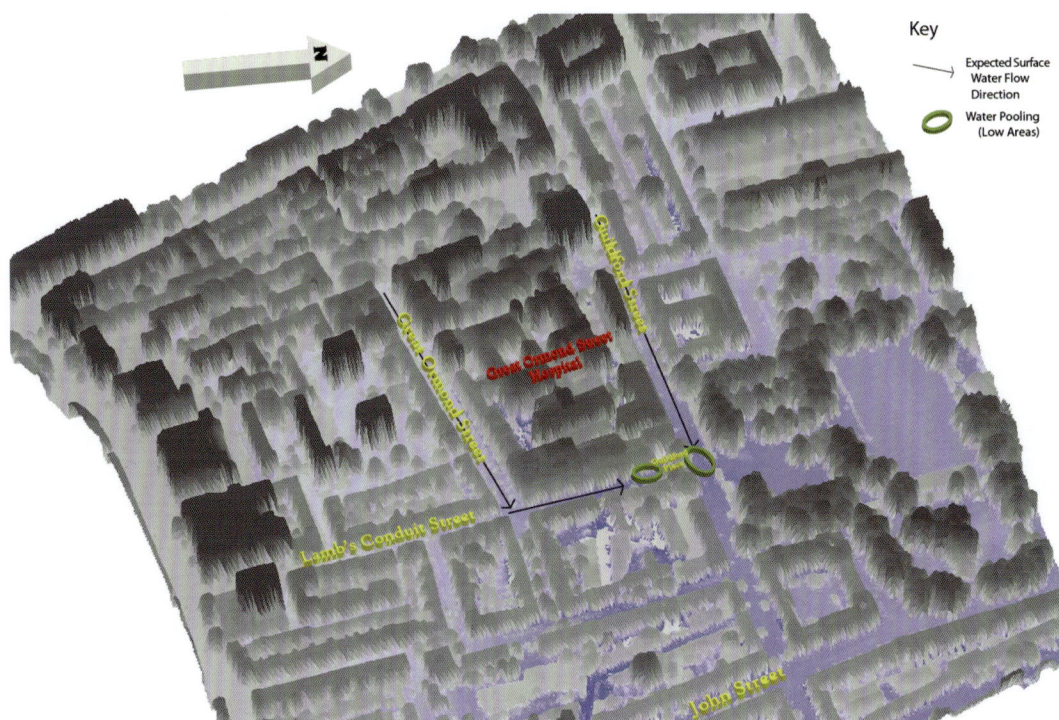

7.10 This image, produced by WSP Built Ecology for the Great Ormond Street Hospital project, shows the topography of the site and the expected direction of surface-water run-off. It was produced using LIDAR laser mapping data.

There were no patient areas in the basement, but it did contain the critical-imaging suites. Although the design had progressed too far for these to be relocated to a less vulnerable part of the building, the team suggested that surface-water flooding in the surrounding area should be monitored as an early indicator of the need to move these critical facilities.

A range of dry-proofing measures was suggested, including avoiding low-level openings close to potential ingress points, installing demountable barriers at the top of a ramp leading from the street down into the basement, and installing non-return valves on drainage. The team recommended a highly detailed assessment of the building, to identify vulnerable areas and to ensure that all potential flooding pathways were comprehensively addressed – a single gap in the defences would render the entire strategy worthless.

WSP Built Ecology also suggested wet-proofing measures such as raising electrical outlet points. In basement areas, however, this strategy may not be effective. Once floodwater overtops a threshold leading to a lower area, that area will fill rapidly to the general flood level – potentially up to or above a basement ceiling, for example.

It was also proposed that a site-wide safe-access and emergency plan should be drawn up and integrated into the water champion role, described above.

Triangle Architects, working on the British Trimmings extra-care scheme, reviewed the landscaping plan of the site with a view to increasing soft areas and therefore its ability to absorb rainfall directly. They found that by using hard surfaces more efficiently – roads that also provide pedestrian access, and footpaths that widen near dwellings to enable outdoor activities – they were able to reduce the area of hard paving by 20%, making a significant difference to surface-water-drainage loads.

They also noted the potential for using a large underground tank for both rainwater harvesting and surface-water attenuation. On their site, the attenuation tank needed to store rainwater from a peak annual storm would be about the same size as 14 days' water supply. Of course, for the attenuation tank to be effective it would need to be emptied before a storm broke, requiring a high level of confidence in weather forecasting.

Groundwater

Groundwater levels may change significantly for a variety of reasons. These may be natural, as with higher rainfall, or man-made – a reduction in water abstraction or the construction of an underground barrier, such as a new flood defence.

A flood defence must be deep enough so that it is not bypassed by river water flowing out underground. But under normal conditions, this may also interrupt natural subsurface water movements in the other direction – the natural groundwater movement is, in effect, dammed by the underground flood defence and the water table rises as a result (see figure 7.10 below). The causes and effects will not necessarily be obvious, and each case needs to be assessed individually.

Few projects were susceptible to high groundwater levels, though one example was the University of Greenwich. Although the basement is above normal groundwater level, it was found to be below the predicted 1-in-1,000-year flood level (which does not take climate change into account). The basement walls were therefore designed to resist groundwater ingress.

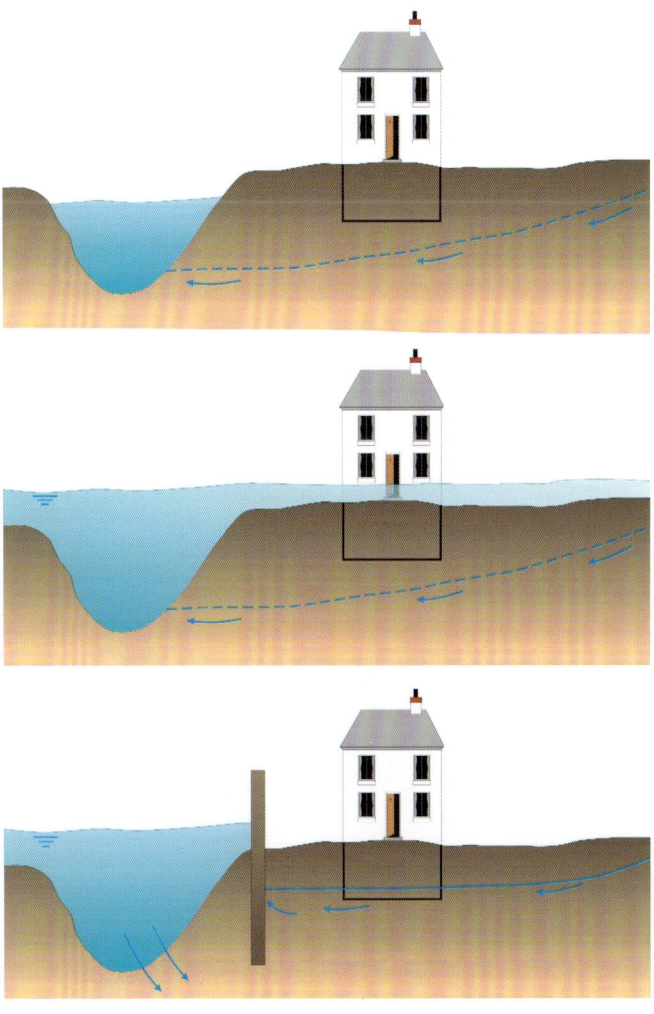

7.11 This diagram, produced by the Environment Agency, shows how flood defences can contribute to rising groundwater levels by blocking flows underground, causing the water table to rise.
Source: Making Space for Water: Groundwater Flooding Records Collation, Monitoring and Risk Assessment (ref HA5) Consolidated Report, Environment Agency, 2007

The work of the TSB projects has highlighted that there are frameworks to enable designers to respond appropriately to the impacts of climate change on rainfall patterns. However, the extent to which these impacts are already taken into account differs from issue to issue.

Increasingly stringent water-conservation standards for new homes and other buildings are being introduced and embedded in regulation. This, of course, presupposes that current targets – such as 80l/person/day for homes – will hold good for a more water-stressed future, and also that the anticipated effects of water-saving devices are actually delivered in practice. Nor is the existing building stock covered well by current water-saving measures.

Our response to river and marine flooding is well coordinated, but the agreed allowances for climate change must be more fully integrated. An integrated response to surface-water flooding is lagging behind somewhat – although there are now clear responsibilities, suitably supported by expertise within the Environment Agency.

Roof and site drainage-design standards are well covered but there is conflicting guidance, and allowances for climate change have yet to be integrated. It is less clear than in the case of surface-water flooding who should be responsible for making this happen.

Overall, the tasks are well defined – but there are gaps in the processes that would enable them to be carried out, and these must be filled.

8 CLIENTS, CONTROL AND COSTS

ENGAGEMENT AND COLLABORATION

We have seen in earlier chapters that a changing climate presents very significant design challenges for the built environment. But it is our clients who ultimately decide what level of change the building they are commissioning should accommodate and how much to spend in order to achieve this.

Adaptation is an emerging issue and, as such, its relevance to different elements of our society – and thus our clients – is becoming clear at different rates. Indeed, several of the Design for Future Climate applications for funding were instigated by clients, demonstrating a growing awareness and the particular relevance of the projected changes to long-term investments such as buildings.

In the majority of cases, it is clear that this was the client and design team's first experience of making concrete decisions about how climate change might affect the design of a building, the relationship between its expected lifespan and the magnitude of projected change, and the cost implications. The requirement for such long-term thinking is in marked contrast to the short-term focus of much current business practice. In some cases, exploring potential impacts for the building project appears to have triggered clients to consider the broader implications for their business, using some of the risk-assessment tools and frameworks described earlier.

As professionals with particular expertise, this is a role in which designers often find themselves: making clients aware of emerging issues and providing them with the information on which to base decisions. A number of teams indicated that they are now offering climate change adaptation workshops as part of their standard service, as a result of experience gained on Design for Future Climate projects.

That said, as adaptation is not yet part of mainstream thinking, it is unlikely that every member of the design and client team will have the same level of knowledge or experience at the start of a project. Bauman Lyons kicked off the Church View project with an introductory workshop, or "Bunker Day", when the design team, client, climate experts, other specialists and stakeholders, including the planning officers, got together for a day of presentations and discussions. This brought the whole team up to speed very rapidly and provided a firm context on which to base the adaptation work. Exercises like this help bring disciplines together. Across the Design for Future Climate projects, a number of teams expressed the value of the wider perspective and cross-fertilisation of ideas that resulted from combining the efforts of academics and professional consultants.

Design is a process of dialogue, an exploration of the fit between a client's expectations and the developing proposals, where designers can raise new issues and test their relevance to the client. The process engenders a common understanding of the implications that issues such as adaptation will hold for the building throughout its lifetime, which those more remote from the dialogue may not appreciate. Triangle Architects noted that those from the client organisation who were directly involved in the project became engaged, interested and enthusiastic, but that the idea of adaptation as a serious issue appeared to gain less traction among those more senior. For this project, there was a view that thinking as far ahead as 2050 and 2080 was too remote and academic, so a 2030 time-frame was introduced into their analysis to provide a more immediate focus for decisions.

PROCUREMENT

The way a building is procured can also have a significant impact on how well its performance in future climates is taken into account at the design stage. Where the client will own and occupy the building, the links between capital cost, cost in use and operational benefits should be relatively seamless. Where those links are broken, the direct input of the building's future users cannot be taken into account and the link between cost and benefit is lost. Designers are already familiar with situations where one party may pay for energy-efficiency improvements while another benefits – not an attractive commercial model. The same is true for adaptation measures.

● ● ●
DESIGN TIME AND TIMING

The Design for Future Climate projects were unusual in that funding was provided specifically to consider adaptation. In the normal course of events, this would either be covered in the client's brief or, depending on their level of knowledge, in the design team's response to that brief – if the issue was raised at all.

Until adaptation is embedded in codes, regulations and standards, this work will require considerable research and thinking time, which costs money and extends the design programme. In this case the additional time was paid for by the TSB, but under normal circumstances the client would need to include an allowance of both time and money for adaptation to be properly considered. This should not be a problem if a client is conversant with adaptation issues, but in today's climate of budget fees and rapid design programmes, a client who is not may take some persuading. Nevertheless, it should be the duty of professionals to raise relevant issues with their clients for them to reject or accept as they see fit.

A number of teams noted the benefits of considering adaptation early on in the design process, preferably at RIBA Work Stage B: Design Brief. On projects where adaptation was only raised when a design was more developed, teams expressed frustration that they could not "unpick" earlier decisions. In some cases, the design had progressed so far that adaptation issues could only be considered for future maintenance cycles or for minor matters of detail rather than the strategic direction of a project.

It should be stressed again that the proper consideration of adaptation issues takes time. Arup, working on the Engineering Graduate School at the University of Sheffield, found that although the project was well resourced, the adaptation analysis fell behind the very rapid design programme simply because of the time needed to model different options, interrogate and understand the results, and iterate improvements. For this research project, they capitalised on the time lag to produce a fundamentally different adaptation-led design to compare with the main project. Clearly, this would not be useful under normal circumstances.

Design teams are thus faced with a dilemma. For modelling purposes, a design must have progressed to a stage where meaningful results can be produced, but not so far that changes are too disruptive. Gale & Snowden's approach of using models made for a previous, similar project in order to draw strategic conclusions was a neat way of getting over this issue, albeit only applicable where such models are available (see figure 8.1 overleaf).

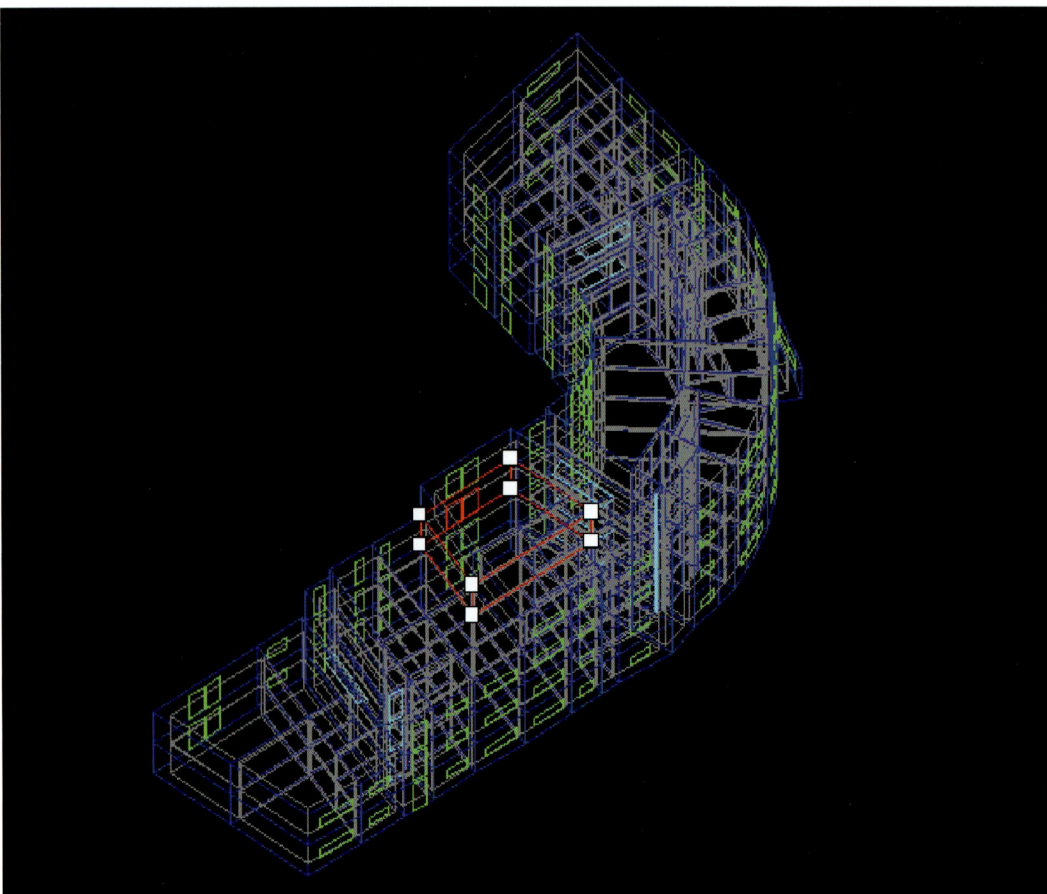

8.1 Gale & Snowden was able to use thermal models developed for a previous, similar, project to assess the impact of climate change on the Extra Care 4 Exeter project at a very early stage in the design process.

For mainstream projects, the assumption is that adaptation would, in due course, become an integrated part of an agreed design service, albeit with some specialist input. The special, research nature of these projects meant that, in some cases, there was a parallel team focused solely on adaptation. This had the virtue that they were able to concentrate on issues that might have been lost in the melee of a fast-moving design. Even when adaptation was covered by the design team themselves, on some projects they opted to have separate, adaptation-focused meetings rather than append the issue to their normal design agenda. Different teams will have different methods of working, but it is important that time is allowed for focused thought, discussion, investigation and testing of options.

PLANNING PERMISSION

As discussed in Chapter 7, consideration of flooding is already an integral part of the planning process – even if allowances for climate change are not yet fully incorporated in flood mapping and relevant guidance, such as PPS 25.

However, obtaining planning consent for adaptation measures that would alter the appearance of a building was flagged as a potential issue, although none of the projects needed to go as far as obtaining formal planning consents specifically for adaptation measures. Certainly, project teams dealing with listed buildings reported considerable nervousness from planning departments.

On the Church View project, for example, there were initial assurances from planners that they would be open-minded about alterations to the listed building, and they participated enthusiastically in the Bunker Day. But when specific interventions were proposed – changing the pattern of windows, for example, or adding shutters – they expressed reluctance to accept them. Had time and budget allowed, Bauman Lyons intended to test the process by submitting a formal planning application, though this was not possible in the end.

Similarly, external alterations to the London School of Hygiene and Tropical Medicine, listed and in the Bloomsbury Conservation Area, appeared to have been almost ruled out of court by the design team in the expectation that the planning department would certainly object.

Perhaps more surprisingly, teams working on projects in relatively "ordinary" locations were also nervous about proposing anything that would alter the appearance of their building. On the British Trimmings extra-care scheme, located in a pleasant though not exceptional suburban area on the outskirts of Leek in Staffordshire, Triangle Architects cited planning as a significant driver towards solutions that did not affect the external appearance of the building. This was perhaps influenced by the fact that the research project began just after the building had been through an "arduous" planning-approval process. Under these circumstances, it clearly would not have been appropriate to disrupt a procurement process that was in full flow, but it is to be hoped that sensible adaptation measures could have been incorporated in the design without making the planning process more difficult.

Planning was not always a barrier, however. Gale & Snowden benefited from involvement at an earlier stage and, possibly, a more amenable planning authority, and the team had time to obtain planning permission for a radically different building form, tuned to future climate, rather than attempting to adapt the design that had received outline planning consent. A similar openness to change was reported by Arup when discussing proposed shading options for the alternative engineering-led design for the University of Sheffield project.

It is still early days for understanding the potential impacts of climate change on our townscape. But there clearly needs to be a more developed discussion about adaptation measures, and which interventions will and will not be acceptable; a discussion that might lead us to re-examine our current priorities. Our buildings have changed in the past – you have only to look at photographs from the pre-air-conditioning era of American or, indeed, British high streets, when the awning was an almost universally adopted device for protecting windows from excessive solar gain. As a society, we can choose to rule out visible passive measures like these to protect our buildings, but the consequence will be increased emissions. At the very least, it should be possible to install removable devices that could make a significant difference to a building's summertime performance without significantly damaging the historic fabric or defacing the townscape.

BUILDING CONTROL

While Building Regulations cover many aspects of building design and construction that will be affected by climate change, they do not yet include allowances for it, relying to a large extent on standards that will need to be systematically updated – a point noted by many of the project teams.

This is a significant task, and one which the government must lead. It is essential to developing a consensus on the reasonable minimum approach that the design and construction community should take to allow for ongoing climate change.

The Design for Future Climate programme has already benefited this process. AECOM is advising the Department for Communities and Local Government on how regulations might better cover overheating, and its experiences on four of the first tranche of projects have provided valuable background. Given the current coalition government's aversion to additional regulation, it remains to be seen how and if regulations will incorporate adaptation. A logical progression might be first to develop consensus and disseminate it as best-practice guidance; then to test that guidance on real projects, much the same way as the Design for Future Climate programme has done; and only then to incorporate a refined and simplified approach in regulation if it becomes clear that the market will not deliver the necessary change to protect future users of our buildings.

MODELLING THERMAL COMFORT

It is not possible to allow for the effects of future climate using current standard compliance tools such as SBEM and SAP (respectively, the BRE's Simplified Building Energy Model and the UK government's Standard Assessment Procedure). Future weather data cannot readily be used with these tools, and criteria to judge the results are not included.

All the teams used dynamic simulation software, able to accept standard-format weather files, to investigate the future performance of their buildings. The choice of tool seemed to have as much to do with office preference or historic accident as with the merits of particular software packages, with different offices in the same organisation sometimes operating different platforms. Teams did flag up limitations in the ability of even these sophisticated packages to model potential adaptation strategies such as ceiling fans or green roofs. In some cases, they used more advanced Computational Fluid Dynamic (CFD) modelling tools to investigate these issues.

Of course, complex dynamic simulation tools are only as good as the people who are using them, and familiarity with a package is essential in interpreting the results. Understanding the – sometimes relatively simplistic – calculation methodology used in the package can be invaluable in identifying rogue results. Similarly, although specialist modellers may know their software inside out, there is no substitute for real-world experience of how buildings behave.

Making sure that the assumptions in the inputs to the model are correct is also essential. If they are wrong, the results will be meaningless. Bauman Lyons noted that one of the key virtues of a fully integrated team was that the architects and facilities manager felt able to discuss, question and refine assumptions made in modelling, normally the preserve of a specialist building physicist isolated from detailed design discussions in which anomalies might be picked up.

A number of teams questioned whether current energy-modelling tools were a valid way of examining summertime comfort, and suggested that they should be calibrated against reality more frequently. Bauman Lyons proposed that building managers should log temperature data and review this every 10 years or so, to spot long-term trends and trigger implementation of progressive adaptation measures.

There is also a danger of being seduced by the apparent precision of these models, and the weather data on which they rely. One must always remember that the building-simulation model itself is not perfect; that the weather data is itself based on the sequential use of a series of models, each one to some extent imperfect; and that, inevitably, errors increase the further into the future you look. Having said that, they are the best tools we have and although we should not expect them to provide "the truth", they should point us in the right direction and enable us to make comparisons between options.

Of course, if changes are made to the design of a building without updating the thermal model, all the effort put into creating an accurate model can be thoroughly compromised. AEDAS, working on the Harris Academy in Purley, suggested that an "as procured" thermal model be constructed, taking account of any changes made during the procurement and construction processes, whether for "value engineering" or other reasons. This would be used to highlight risks for building managers and occupiers, as well as specific changes that might adversely affect the design intent.

Is the wider adoption of Building Information Modelling, with its promise of seamless interoperability across the design and construction team, all working through a master model, the answer? The distance that the industry needs to travel towards this holy grail is illustrated by the experience of Arup's sustainability specialist, working on the Church View project. The pre-existing dynamic simulation model was not detailed enough to analyse particular rooms in sufficient detail in order to assess adaptation strategies. Initially, attempts were made to transfer information directly from the architect's ArchiCAD model, but, paradoxically, the complexity of simplifying this data-set was so great that in the end it was quicker simply to rebuild the environmental model.

The dual use of project design information is not just a technical issue. There are contractual barriers, as AEDAS found working on the Harris Academy. Here a separate "research" model had to be constructed because of a concern over intellectual property rights over the "contract" model produced by the building services engineers, and an understandable nervousness from those who had built the model to allow others to develop or modify it. The research model was used to test risks and potential adaptations, which were then resimulated in the contract model by the building services engineers. These are real barriers to efficient working and, again, have profound implications for a BIM future.

Arup also noted that while it was quite simple to swap weather files when modelling a building, it was a much greater challenge to deal with the huge amounts of data produced as a result and to compile it in a form from which conclusions could be drawn. Very rigorous data-handling protocols were required to keep track of all the model iterations, and the team noted that one of their key achievements was the development of template spreadsheets that allowed raw dynamic simulation modelling outputs (typically 200MB Excel files) to be organised and visualised in a way that enabled the team to make sense of them.

Architects are very familiar with the use of environmental modelling tools to improve the energy efficiency of their buildings, but few have wanted to get to grips with the modeling process in any depth. First-hand experience of modelling on the Design for Future Climate projects, however, seems to have had a profound effect on the attitude of the architects involved. They have seen how computational tools can help them understand the direct relationship between the fundamentals of architecture: natural light, the movement of air and the effect of materials on the environmental experience of a building's occupants. Some practices already had in-house modelling capability, but for architects like Bauman Lyons it has been the trigger to invest in building-simulation software to enable them to use and control environmental modelling on their projects from an early stage.

● ● ●
ADAPTATION STEP-BY-STEP

Most project teams recognised that adaptation is an ongoing process. Rather than incorporating everything a building will need by 2080 immediately, measures can be applied as and when required, and when there is greater certainty about the speed of change. They displayed this information in different ways, a selection of which are shown below.

Hoare Lea summarised recommendations for the Oxford University Press in a very simple table, setting out a provisional programme for interventions that was organised to suit anticipated maintenance and replacement periods.

ITEM	ADAPTATION MEASURE PROPOSED	APPLICABLE YEARS				REASONS FOR SELECTION
		NOW	2020	2040	2080	
1	Alteration to the roof design on the west end of the building to allow for a future plant mezzanine.	✓		✓		The chillers are not predicted to be insufficient to cope with the load until after 2040. However the changes to the roof design to allow for additional plant should be included as part of the current structural design to ensure the structure is capable of holding the extra weight.
2	The addition of thermal mass to the top floor ceiling of the existing building to increase the thermal mass of the building to help mitigate overheating.	✓				This option would not only help militate against future overheating, but would also reduce cooling energy use in today's climate. Therefore it is recommended that this is implemented now.
3	To help mitigate the risk of future overheating the exposed concrete ceilings could be coffered with blanked pipes contained within. This is to allow a future chilled ceiling system to be installed with minimal impact to the structure.	✓				As above, this would help reduce electrical use today so is applicable to today's conditions.
4	The inclusion of a 'knock-out panel' next to the south-east riser to allow for future additional services.	✓				As the cooling is not predicted to be required to be increased until sometime after 2040 the additional services would not be needed until then. However the knock-out panels would need to be installed today to avoid unnecessary structural work.
5	The boilers to be more modularised to allow for easier adaptation to future changes in climate.			✓		The boilers will be sufficient to cope with the reduction in heating demand until at least 2040.
6	Three additional drinking points to be provided in the quad and on the roof.	✓				Heat waves can occur at any time so this adaptation measure is recommended for implementation now.

● ● ● **8.2** Hoare Lea used this table to link adaptation measures to a proposed timescale for its client at Oxford University Press.

WSP Built Ecology followed a similar approach at Trowbridge County Hall, including guidance on monitoring and the factors that should be taken into account when deciding whether or not to implement a strategy.

RISK	ACTION / STRATEGY			
	Current refurbishment	Next potential upgrade (2030) – Building Services	Following upgrade (2050) – Building fabric and facade	Following upgrade (2080)
Uncomfortable internal environment	Expanded landscaping to provide additional trees. Provide additional insulation.	Based on current 2030 performance and projected future performance, seek to incorporate opportunities linked to the upgrade works (in 2030 is likely to be limited to HVAC).	Based on current 2050 performance and projected future performance, seek to incorporate opportunities linked to the upgrade works (in 2050 is likely to extend to facade and building fabric).	Reassess thermal comfort in key occupied spaces using current 2080 weather data.
		Additional HVAC capacity.	Consider the following in the context of the cost-benefit analysis: upgrade glazing performance specification, night purge, exposed thermal mass, external shading, glazing film.	
Uncomfortable external environment	Expanded landscaping to provide additional trees.	Reassess thermal comfort in key external spaces using current weather data.		
Structural instability from shrink and heave	Undertake a soil pit test to establish the risk profile.	Monitor the soil subsidence of surrounding residential buildings for evidence of subsidence. These buildings would be more susceptible and would provide advanced warning of any issues.		

● ● ● 8.3 WSP Built Ecology summarised adaptation strategies, a potential timetable and the factors that should be taken into account in the future in a table for its client at County Hall, Trowbridge.

At Great Ormond Street Hospital, the WSP Built Ecology team took a slightly different approach, setting out adaptation measures with indicative costs and assigning them a level of priority from 1 to 4, for consideration immediately and over the next 20 years (see figure 8.4 opposite).

● ● ● (opposite)

8.4 Table of potential adaptation strategies produced by WSP Built Ecology for Great Ormond Street Hospital.

*Priority 1: Initiative is recommended to be adopted now either because there is an obvious design advantage or because it is significantly cheaper to do so

Priority 2: Initiative should be considered for adoption in the next 20 years or as part of the next building upgrade

Priority 3: Initiative should be considered for adoption after 20 years or as observed climate conditions may require

Priority 4: Initiative is not part of the climate change adaptation strategy at this time

TOPIC	OPPORTUNITY	IMPACT	COST	RECOMMENDED
Water resources availability	Water resources strategy (WRS)	More sustainable use of water resources, reduced expenditure on water and flood-risk mitigation through strategic management of water.	Requires an internal 'water champion' to lead and implement. Provided funding can be secured (eg Aquafund) this initative should be little or no cost to GOSH.	Priority 1*
	Water efficient appliances	More sustainable use of water resources, reduced expenditure on water.	No additional cost: Included in proposed design of Phase 2B building. Other parts of the hospital: short payback period for WCs and urinals, slightly longer for water-efficient taps. Low-cost options available: cistern displacement devices and flow restrictors.	Priority 1 Priority 1, 2, 3 for other parts of the hospital
	Education and awareness	More sustainable use of water resources, reduced expenditure on water and behavioural change of staff and patients (increased social capital).	No cost if implemented through Thames Water.	Priority 1, 2, 3
	Rainwater harvesting	More sustainable use of water resources, reduced expenditure on water. Reduced discharge into the public sewer network hence flood-risk mitigation.	£353,000 – outrun cost. £323,294 – life-cycle replacement & maintenance costs.	Priority 1
	Borehole	Reduced expenditure on water and additional water-supply redundancy.	Cost estimate (for 50m³/day abstraction): Collection of hydraulic data and interpretation – £750 Borehole drilling and installation chamber and pumping equipment – £6–8,000 Hydrogeology consultants professional fees – £5–7,500 Licence application – £1,500 assuming no water features survey required Annual licence costs – TBC depending on quantity required Annual monitoring and maintenance costs – TBC	Priority 1
Surface water flood risk	Relocation of assets	Reduced impact of flooding through removal of vulnerable assets from areas at risk.	Potentially expensive depending on the amount/nature of assets to be relocated.	Priority 4
	Dry and wet waterproof measures	Reduced impact of flooding through protection of buildings/assets at risk.	£172,000 – basement tanking. £185,000 – improved drainage blanket.	Priority 4
	Safe access and emergency planning	Mitigate risk to people in case of flooding through appropriate emergency procedures.	Low cost limited to the time spent in the emergency plan preparation, dissemination and review. To be implemented as part of the WRS.	Priority 1, 2, 3
	Green roofs	Reduced discharge rates (ie flood-risk mitigation). Longer lifespan for the roof, better insulation, improved ecology, air quality and potential social benefit.	£569,000 – outrun cost. £579,941 – life-cycle replacement & maintenance cost.	Priority 1

At the Harris Academy, AEDAS combined the two approaches above by setting out a decade-by-decade chart of potential interventions, linked to refurbishment cycles with associated costs.

TIMESCALE	CURRENT	2020	2030	2040
1955 block	Interior refurbished	Interior refurbished Furniture and fittings replaced IT upgrade. Windows/roof replaced	(Major refurbishment/replacement)	Interior refurbished Furniture and fittings replaced
2002 block	Interior refurbished	Interior refurbished Furniture and fittings replaced IT upgrade Windows/roof replaced	Interior refurbished Exterior refurbished	Interior refurbished Furniture and fittings replaced
2007 block		Interior refurbished Furniture and fittings replaced IT upgrade	Interior refurbished Exterior refurbished	Interior refurbished Furniture and fittings replaced Services replaced
2012 block		Interior refurbished Furniture and fittings replaced IT upgrade	Interior refurbished Exterior refurbished	Interior refurbished Furniture and fittings replaced Render and roof covering requires surface repair Services replaced

Designing for comfort	Measures	Methods	Costs	Methods	Costs	Methods	Costs	Methods
	Increased daytime ventilaltion	Installation of extended restrictors	Approx £500 per unit					Replacement windows with increased free area
	Increased night time cooling	Increased actuated openings	Minimal					Replacement windows with increased free area
	Increased thermal mass	Concrete and lining board density	50% inc' to boards	PCM to interior furniture	TBC			PCM to interior furniture
	Reduce solar exposure	Improve design G-value to 0.32	5% to initial glazing	Option to add internal film to achieve G-value of 0.32 or 0.25	£30/m² (today's prices)	Increase shades if appropriate external planting		Cladding replacement and upgrade, increasing G-value
	Reflective materials selected			Refurbish, patch and repaint	Labour and materials			Refurbish, patch and repaint
	Insect screens	Retrofit if required	Retrofit if required, £10/m²					
Keeping cool – building design	Cooling			Green roof retrofit to refurbished block	Approx £150/m²			Increase in individual units or look at cooling systems
Keeping cool – external spaces	Reduce direct solar exposure	Courtyard design and PV covering roof.		Landscaping and community facility		Pergolas to parking, water features etc		
Keeping warm	Reducing losses	Thermal bridging allowance included. Artightness improved		Interior drafts filled		Replace heat source, fabric performance upgrade		

● ● ● **8.5** Two tables produced by AEDAS for its client at the Harris Academy, linking potential adaptation strategies to maintenance cycles and costs, in this case for thermal comfort.

2050	2060	2070	2080
Interior refurbished	Interior refurbished Furniture and fittings replaced Windows/roof replaced	Interior refurbished	Interior refurbished Furniture and fittings replaced
Interior refurbished Windows/roof replaced	Interior refurbished Furniture and fittings replaced IT upgrade Exterior refurbished	Interior refurbished	Interior refurbished Furniture and fittings replaced
Interior refurbished Exterior refurbished	Interior refurbished	Interior refurbished Furniture and fittings replaced Exterior refurbished	Interior refurbished Services replaced
Interior refurbished External cladding and glazing replaced Exterior refurbished	Interior refurbished	Interior refurbished Exterior refurbished	Services replaced

Costs	Methods	Costs	Methods	Costs	Methods	Costs
15% increase window cost	Internal ceiling fans	£150 per unit, 4 per class	Investigate solar extract to stack			
15% increase window cost						
			PCM to interior furniture			
5% increase to window cost						
Labour and materials						
					Solar cooling systems?	
£8000 per unit (today's prices)						

At Church View, Bauman Lyons adopted both a step-by-step and a room-by-room approach, which reflected both the variety of spaces in the building that they were refurbishing and its multi-tenancy nature. The team identified eight "archetype" rooms spanning the variety of sizes, orientations and types of construction, and analysed each of these in detail. The results of their analysis were comprehensive and presented in an extremely engaging and informative way, described in more detail in Chapter 5 (see pages 107–110). Their summary table shows how different interventions might be carried out in each space over time.

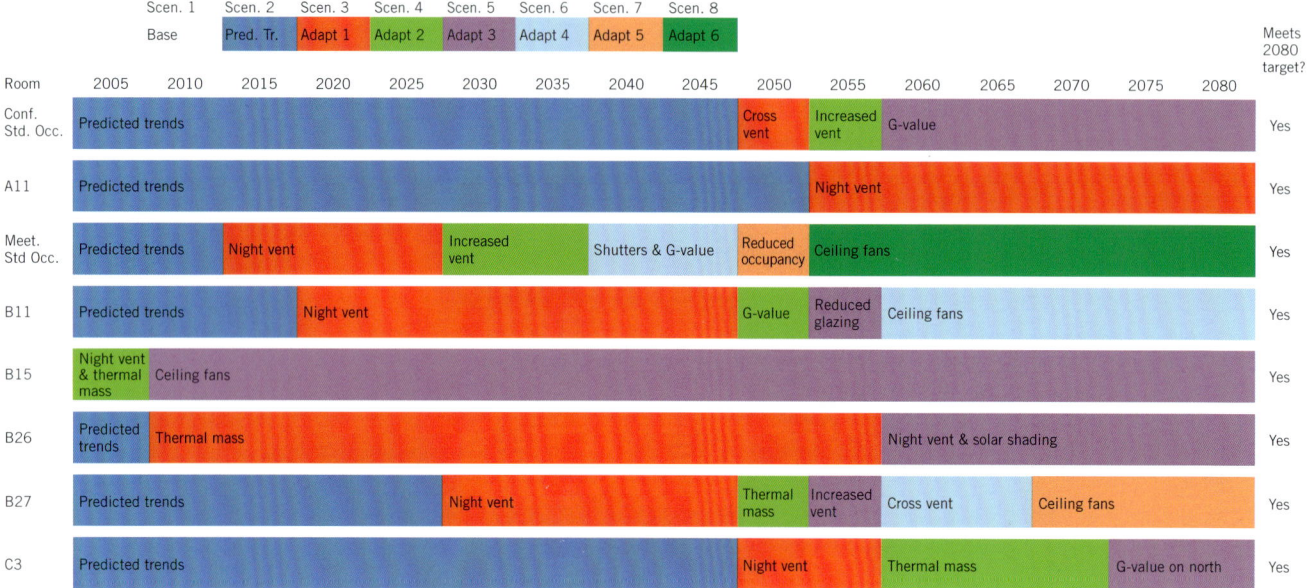

8.6 Bauman Lyons summarised adaptation strategies for all sample rooms at the Church View project in Doncaster in this bar chart.

<h2>● ● ●
COSTS AND BENEFITS</h2>

Cost information was provided by each of the projects, albeit in a variety of forms that reflected different approaches. There is a wealth of specific cost information in the project reports, but evaluating the corresponding benefits of adaptation interventions proved more difficult. Financial benefits were relatively simple to establish for some interventions – those which resulted in a direct reduction in cooling costs, for example – but others could be described only in qualitative terms.

For the Church View project, there was a comprehensive costing exercise based on five alternative options. This indicated that, over 70 years, the whole-life cost of implementing all intervention packages would be around 20% less than installing enhanced cooling to the base-case building. The options did include a number of assumptions about running costs and reductions in income due to overheating, illustrating the difficulty of looking so far into a changing future. Nevertheless, the study gave the client some comfort as to the cost-effectiveness of the interventions. In reality, because a sequential approach has been taken, interventions can be reassessed using up-to-date information when it is appropriate to implement them.

In contrast to this long-term view, the client for the British Trimmings extra-care scheme felt that beyond 30 years, forecasting financial implications through life-cycle costing techniques was subject to too many variables to be reliable, regardless of any uncertainties about climate change. Meanwhile, the Extra Care 4 Exeter project illustrated that a holistic design strategy, such as Passivhaus, offers additional financial benefits, aside from those that result from tackling the specific adaptation issue – in this case, overheating. They took a long-term comprehensive view on costings, producing bar charts to show the comparative energy costs of a Building Regulations-standard building and one built to Passivhaus standards. The adapted building does avoid significant cooling costs, but savings in heating costs are significantly greater. Attributing a cost benefit under these circumstances is not straightforward.

Energy-related costs for an extra-care facility built to 2010 Building Regulations

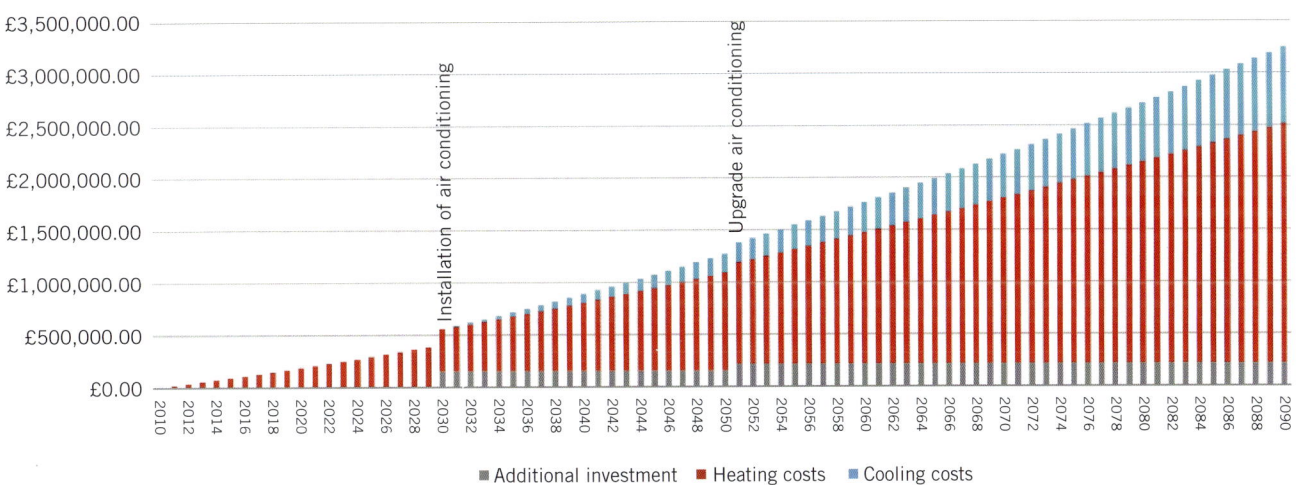

Energy-related costs for an extra-care facility adapted using strategies outlined in report

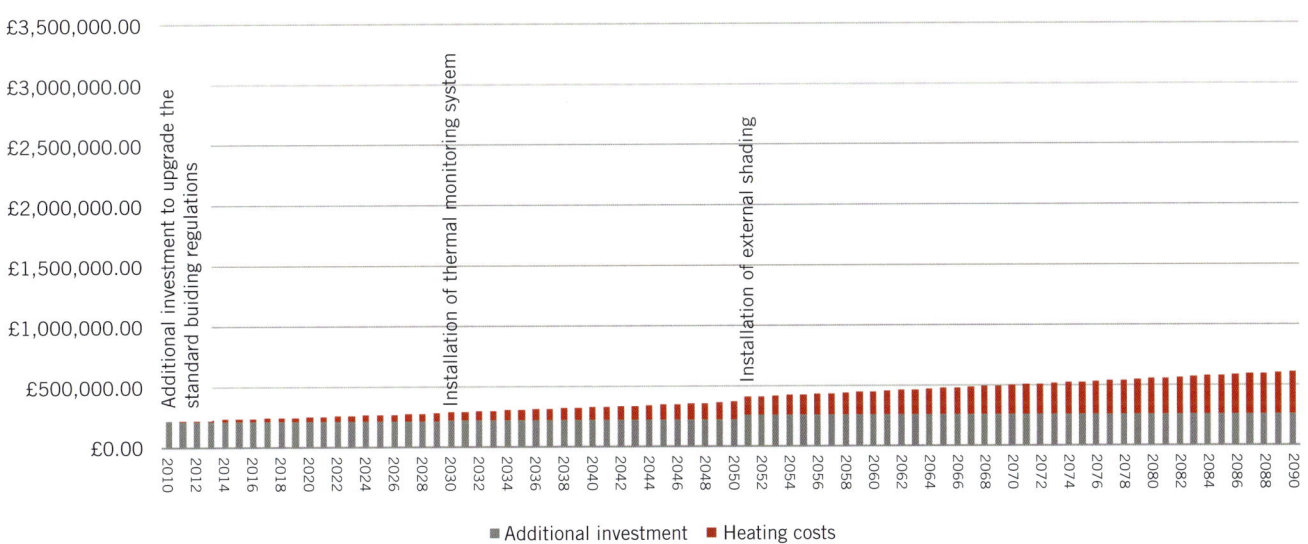

8.7 Gale & Snowden produced these graphs for the Extra Care 4 Exeter project to show how the costs of heating and cooling might change over time, for a building that meets 2010 Building Regulations (above), and for one incorporating their recommended adaptation strategies (below).

In comparison, it is relatively simple to cost the direct benefits of adaptation strategies for buildings that require comfort cooling in today's conditions. Here, it is assumed that the building will remain comfortable but use more energy in order to achieve the required level of comfort. The team can simply test the intervention in their model, calculate the energy savings and make some relatively simple assumptions about future energy costs.

This information can be expressed as annual cost savings: the approach taken for the new Admiral Headquarters in Cardiff, illustrating the effectiveness of zero-cost behavioural options.

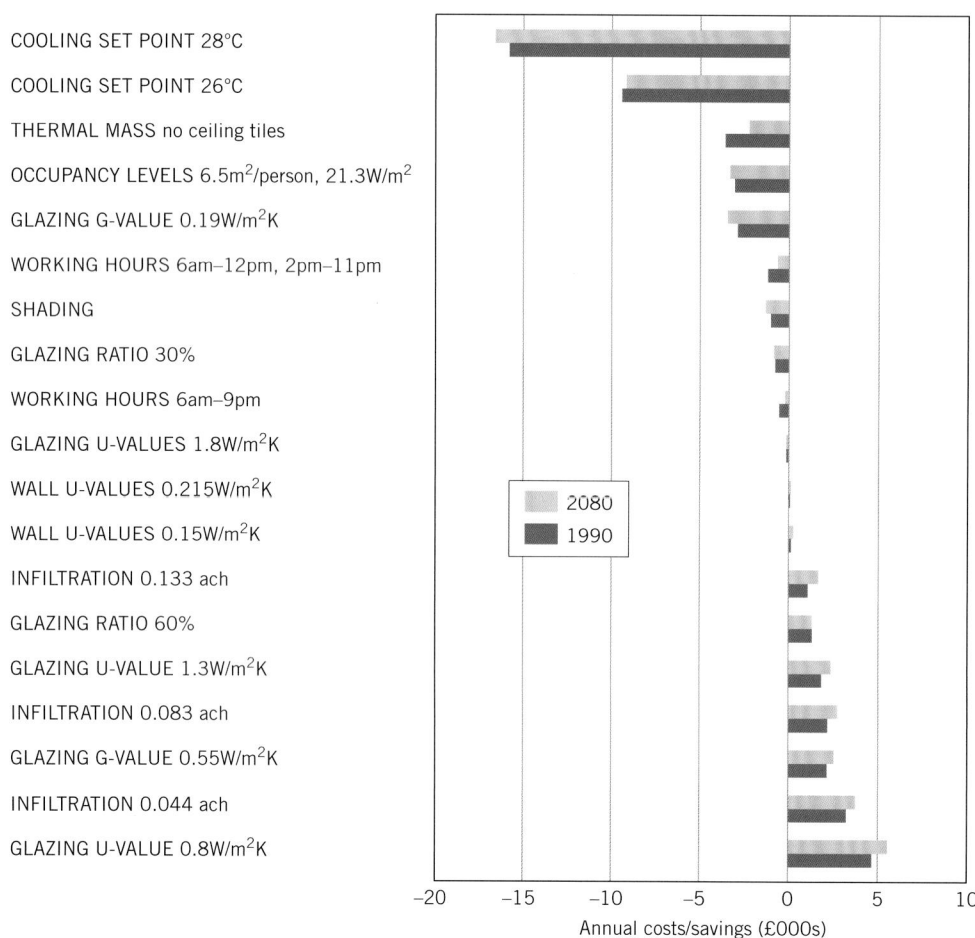

8.8 AECOM produced this graph to show the annual costs and savings associated with proposed adaptation strategies for the Admiral Insurance offices project in Cardiff.

On the EBI Hub, AECOM chose to present information for four adaptation options in a simple table that included cost- and energy-payback periods. This demonstrated that one of the most cost-effective interventions would be to install more efficient lighting and equipment, saving money now and improving conditions in the future.

		UNIT	PRESENT	2050s	2080s
Baseline building	Gas use	MWh/year	302.45	229.70	199.24
	Electricity use	MWh/year	622.42	649.10	665.32
	Gas cost	£/year	£8,771	£6,661	£5,778
	Electricity cost	£/year	£62,305	£64,975	£66,599
	Total cost	£/year	£71,076	£71,636	£72,377
Measure 1: Glazing	Gas use saving	MWh/year	-7	-5	-4
	Electricity use saving	MWh/year	30	33	32
	Cost saving	£/year	£2,779	£3,128	£3,111
	Simple payback	year	123	109	110
Measure 2: Cooling set-point change	Gas use saving	MWh/year	3	2	2
	Electricity use saving	MWh/year	8	8	8
	Cost saving	£/year	£893	£880	£873
	Simple payback	year	0	0	0
Measure 3: Lighting & equipment improvement	Gas use saving	MWh/year	-5	-3	-2
	Electricity use saving	MWh/year	202	203	204
	Total cost	£/year	£50,963	£51,352	£52,048
	Cost saving	£/year	£20,113	£20,284	£20,329
	Simple payback	year	4	4	4
Fully adapted building	Gas use saving	MWh/year	-8	-5	-3
	Electricity use saving	MWh/year	232	236	236
	Total cost	£/year	£48,107	£48,146	£48,818
	Cost saving	£/year	£22,969	£23,491	£23,559
	Simple payback	year	18	18	18

● ● ● 8.9 A table comparing the cost savings of four adaptation measures produced by AECOM for the EBI Hub project.

The lack of correlation between the capital cost of an intervention and its effectiveness was neatly illustrated on the University of Sheffield project by Arup's plot showing the cost of a measure against the percentage reduction in hours over the Adaptive Comfort Threshold that it delivered.

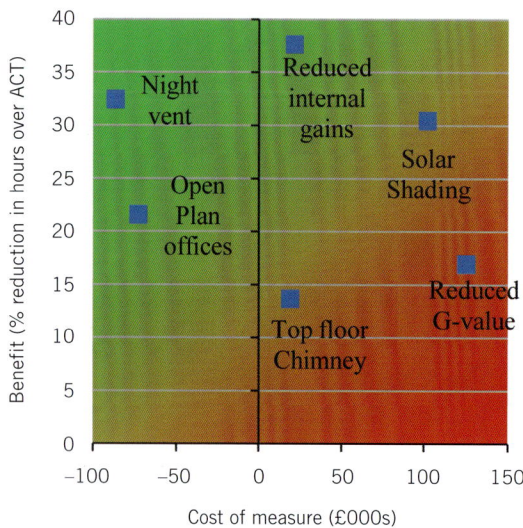

8.10 On the University of Sheffield project, Arup found there was little correlation between the cost of an adaptation measure and its effectiveness at reducing the percentage of occupied hours when the building would overheat.

AEDAS made the interesting point that there is also no correlation between the design costs involved in establishing the effectiveness of an intervention and its capital cost. Some very low-cost interventions require considerable design time, which is not usually taken into account in cost–benefit analysis of this kind. Because they were looking at very detailed interventions – such as increasing the free area of window ventilation by careful specification of actuators – they found it difficult to unpick the complex relationship between different components to produce a straightforward "shopping list" of interventions.

For example, changing from built-in concealed window actuators to cheaper face-fixed ones that provided a greater free area had a knock-on effect by adding complexity to insect-mesh screens, which then had to be adapted to fit around them.

The 100 City Road team raised the issue that the adoption of highly effective adaptation measures such as thin-client IT systems, which could affect sizing of mechanical plant and ventilation strategies, would depend on the future occupants of the building. They also noted that the additional cost of some adaptation measures might increase capital costs for a building's developer while offering savings to building users. It is difficult enough to convince developers to include energy-saving strategies that have an immediate benefit for their potential tenants, even on the basis that reduced running costs will allow them to charge a higher rent. It is another step in the logic to convince them to include adaptation measures that may not come into their own until more extreme conditions, anticipated at some point in the future, apply. To assist in these discussions with their client, they listed adaptations using a traffic-light coding system, ranking their cost, energy and comfort impacts as high, medium or low.

	ADAPTATION STRATEGY	IMPACT OF ADAPTATION ON:			DETAILS
		COST	ENERGY	COMFORT	
1	Extending comfort design criteria	Low +	High +	Medium	Significant saving can be achieved with no real additional capital cost. Up to 20% energy saving can be realised for UKCP09 High emissions 90th percentile climate scenario.
2	Limiting solar gains	Medium	Low –	High +	Shading devices are effective at reducing the peak cooling load of the building. The reduction in energy usage of the building is minimal when compared to the additional heating and lighting energy that could be saved from daylighting and winter passive heating. Retractable shading devices with good automatic controls would provide some energy saving, but detailed analysis against the reduction in daylighting will be required.
3	Limiting internal lighting gains	Low +	High +	Medium	By providing general office background lighting at 200 lux supplemented with task lighting, energy saving is approximately 15% regardless of which weather files were used.
4	Limiting internal equipment gains/use of thin-client system	Low +	High +	Medium	Reducing equipment load by 'thin-client' technology or similar reduces electrical consumption significantly. Cooling energy is reduced on average by 10%. Recovering heat from the equipment gains in the server room should be considered as up to 30% of the office heating can be recovered.
5	Optimising building fabric	Medium	Low –	Low –	Improving the building fabrics reduces the heating load to the office significantly. However, in all the cases simulated, additional cooling energy was required due to more heat being trapped in the building by the improved thermal envelope. In the 2080 simulation, there was actually an increase in carbon emissions due to warmer winters resulting in less heat loss while the cooling energy demand has increased for the building.
6	Optimising thermal mass	High –	Low –	Low –	Effect of cooling load reduction is minimal.
7	Optimising room height & air distribution system	Medium	Medium	Medium	Reducing the floor-to-ceiling height to 3m improved thermal comfort due to reduced solar gains and increased radiant effect from the chilled ceiling resulting in cooler resultant temperatures. Stratifications were observed for all room heights tested and no significant difference was observed by increasing the ceiling height.
8	Night-time cooling	High –	High +	Medium	Saving in cooling load with night-time ventilation strategy is significant for both building with or without embedded pipes. Cooling load in general is reduced by up to 30% when night-time conditions are favourable. Night-time ventilation should be provided efficiently and cost of automatic openings should be reviewed against the saving made from running cost.
9	Mixed-mode ventilation	Medium	High +	Medium	An energy saving of 5–10% can be achieved by naturally ventilating the 4.5m perimeter zones. This is based on windows opening when the outdoor temperature is between 20–25°C and the internal summer design condition at 26°C.
10	Cross ventilation and low-energy air supply	Medium	High +	Medium	Analysis indicated that single-sided ventilation on still, warm days is unlikely to provide effective ventilation. Wind-driven cross ventilation would provide higher ventilation rates. The floorplate of 100 City Road is too deep to ensure good air quality in the central and downwind sides of the office against conventional design criteria. However, cross ventilation could be effectively used for night-ventilation cooling and use of ceiling fans would further enhance these benefits.

● ● ● **8.11** The team working on 100 City Road used this table to summarise the impact of each of the adaptation strategies investigated, where '+' represents a positive impact and '-' represents a negative impact (eg high cost is negative, but high energy saving is positive).

Building for a changing climate is unfamiliar territory for both designers and clients. Those new to the field may be daunted by the potentially bewildering range of available data, which simultaneously acknowledges the certainty that change is under way while also reflecting uncertainty over the rate and extent of that change. The issue cannot be ignored, however, and decisions must be taken on a project-by-project basis until there is a consensus to narrow down the range of options that must be considered and to embed an interpretation of the climate data into the standards, codes and regulations with which the industry is already familiar. The outputs of the Design for Future Climate projects are an excellent first step towards building that consensus.

Best value will be obtained by considering the implications of climate change at the start of the design process. This provides an opportunity to produce innovative, coherent design responses that temper potential impacts through intelligent, passive design. The alternative is a "bolt-on" philosophy that inevitably adds cost, and which, given the long time-frames and layers of uncertainty involved, is difficult to justify commercially.

9 WHAT NOW?

Analysing the extensive range of outputs from the first tranche of Design for Future Climate projects, and discussing the work with many of the people involved, has been a fascinating and rewarding experience. The funding provided by the Technology Strategy Board has enabled design teams and their clients, working with climate experts and academics, to consider the implications of a changing climate on live projects in far more depth than would usually be possible in the cut and thrust of day-to-day project work.

For many of the teams this will have been their first encounter with the design implications of adaptation, and it has been very encouraging to see how the knowledge and expertise built up over a relatively short period has produced some real insights into the issues facing the industry.

In this respect, the projects have acted as pathfinders. They have been exploring currently available tools and information in a live project environment and starting to develop robust approaches to new challenges, ready for wider adoption by the industry. They have also been flagging up the inconsistencies and gaps that should be filled as we get to grips with adaptation. This model could perhaps be followed for testing future enhancements to the Building Regulations before they are implemented.

Valuable lessons have also been learned by the clients involved – not only in considering the implications for their buildings, but for their businesses too.

Hopefully, this book provides a useful summary of the key issues that have emerged. However, the final project reports are freely available to download at the TSB_connect website[21] and provide a rich source for those wishing to find out more about particular projects or explore specific issues or methodologies in more detail.

It would be no exaggeration to say that for many of the participants, the opportunity to focus on adaptation has fundamentally changed their approach to design. In many cases consideration of climate change has become an integral part of their design service, and they have begun raising these issues with their clients as a matter of course.

One important realisation has been that the "future" is already here, and that the standard weather files on which they have previously relied are effectively out of date, being based on averages of historical observations that, inevitably, lag behind the changes underway.

Having had the opportunity to stand back from the "drawing board" and review projects through the lens of a changing climate, teams also identified some serious shortcomings and these have prompted new design ideas.

Detailed discussions with engineers and building physicists have also changed some architects' perceptions of environmental simulation models. For many, this has been their first chance to engage fully with the inputs and outputs of these models, and they have recognised how useful they can be in informing fundamental decisions about architecture. Some have even brought such software in-house, so that it is no longer a remote energy performance tool left in the hands of specialists but under their direct control. This engagement with building physics is to be welcomed, and hopefully heralds a renewed interest in environmental design – though it should enhance collaboration between disciplines rather than replace the input of a good environmental engineer.

The broader implications of climate change were not generally raised. This was an entirely reasonable omission in the context of work that was focused on live projects, and considering the limitations of the available information (as well as reflecting an element of pragmatism in the face of the huge levels of uncertainty in the longer term). However, it is worth raising one of these implications here because it illustrates the connection between the mitigation and adaptation agendas.

Based on their assessment of trends in global emissions and clients' vulnerability to change, the majority of teams opted to base their studies on the "High" emissions scenario. Several explored strategies to deal with the more extreme end of the probabilistic range for this scenario, demonstrating that it is possible to design buildings that can maintain comfortable conditions even under such circumstances. On the other hand, policy makers need to bear in mind that the consequences of global warming of this magnitude would be catastrophic elsewhere in the world, and would be likely to fundamentally change socio-economic activity in this country. We should not allow ourselves to be lulled into a false sense of security by a theoretical ability to adapt individual buildings, and assume that the mitigation agenda to reduce emissions can be relaxed. Mitigation and adaptation are inextricably linked: the more effective we are at mitigation, the less we will need to adapt. The consequences of not achieving our mitigation targets will be far wider in scope than just uncomfortable buildings.

DATA AND TOOLS

Teams were able to draw on a range of future-climate weather files to predict comfort performance, "translated" from raw climate information into a standard format that can be used in building-simulation software to predict performance throughout the century – an admirable example of how information about future climate can be embedded in normal design practice.

When the Design for Future Climate competition was launched in 2010, the only source of future weather data was that developed by CIBSE in collaboration with UKCIP and Arup (based on the 2002 climate projections). As the projects got under way, new probabilistic future weather files became available as outputs from the EPSRC-funded ARCC portfolio of research projects (PROMETHUEUS and COPSE), produced using the UKCP09 Weather Generator. CIBSE, again working with Arup, are in the process of producing new probabilistic weather files, a trial version of which was used on one of the Arup projects here. These will be based on a similar morphing technique to that used for their previous future weather files. There are pros and cons for morphing as compared with Weather Generator-based approaches, particularly concerning how well the resulting data reflects extreme events and whether it is possible to produce tailored data for locations that do not have adequate observed records, as well as initial concerns about anomalous results produced by the Weather Generator.

The availability of a selection of future weather files is a significant step forward, but it also threatens to add complexity where clarity is needed. It offers individual design teams and their clients the ability to choose a weather data-set and decide which emissions scenario, time-frame and probability level best reflects perceived vulnerabilities and the anticipated lifespan of a building. But it also makes comparison between buildings almost impossible, and many will be daunted by multiple layers of decision-making about unfamiliar issues.

Our understanding of how to design buildings that will remain comfortable as our climate warms is obviously in its initial phase. A number of teams pointed out that proper consideration of adaptation consumed much time – the scarcest resource available to designers. They needed to construct models in sufficient detail to understand the implications of change, interrogate the results and then draw conclusions to be fed into further iterations. For this issue to be taken up by the mainstream, the field of choice needs to be narrowed and guidance established as to the "reasonable" conditions against which a building should be tested. Many of the teams noted that this needs to happen as soon as practically possible, and should be based on a consensus established by early adopters such as the 50 or so projects so far included in the Design for Future Climate initiative.

Issues associated with river and tidal flooding were found to be already well embedded in our planning system, and though current flood designations and mapping do not yet include allowances for climate change, there are recommended national precautionary allowances for increases in peak rainfall intensity, peak river flow, offshore wind speed and extreme wave height throughout the century. This provided a useful starting point for the analysis of future impacts.

Some teams delved deeper into the issues, developing potential enhancements to standard flood-risk modelling in order to better take account of future extreme events. As with comfort issues, appropriate allowances for climate change should be integrated into flood mapping and guidance as soon as reasonably possible. The fact that the Environment Agency has overall responsibility and is the acknowledged centre of expertise should make the process of developing a consensus relatively simple.

Information and guidance about surface-water flooding appears to be a step behind river and tidal flooding, but the structure of Lead Local Flood Authorities is now in place to support a similar process of incorporating allowances for climate change.

Improved water conservation is already well covered under Building Regulations and, as all of the projects were targeting high BREEAM ratings, most practical water-saving approaches had been already included and teams did not identify further useful measures. However, where water-conservation strategies depend on rainwater collection on site, the methodology for sizing storage capacity does not yet take account of climate change. Teams working on tight urban sites noted the difficulties of providing adequate storage, particularly if tank sizes need to increase, and suggested that this might better be dealt with at a larger scale rather than on an individual site-by-site basis. This raises the question of how the total costs, embodied energy and maintenance burden that accompany small-scale rainwater collection compare with larger-scale systems.

Teams had greater difficulty in taking climate change into account for other aspects of construction, including roof drainage, structural design, and materials and detailing standards. For these, the industry relies very heavily on standards and codes that do not as yet include appropriate allowances. The teams identified some of the inherent difficulties in developing such allowances, in some cases proposing interesting methodologies to compensate for a lack of hard information.

One of the difficulties is that UKCP09 does not provide full information for some key factors, notably wind data and storm intensity, because there is insufficient correlation between climate models to allow robust probabilistic projections. Another is that where information is available, it is not in a form that can be directly used in building design. For example, future rainfall projections are available but only in the form of averages – typically daily or hourly amounts rather than figures reflecting the brief intense events on which the design of roof-drainage systems is based. Again, there needs to be consensus on how to update the standards with suitably testing data to allow for future climate.

Existing standards and codes provide a clear framework, into which appropriate allowances for climate change could be slotted. But it is less clear how this might happen for construction factors, in comparison with comfort, flooding and water shortage, because there seems to be little correlation between centres of expertise and responsibility for systematically updating standards.

● ● ●
COSTS AND MARKET DRIVERS

The costs associated with implementing adaptation measures were reported by project teams, but as the nature of the projects varied quite widely it is difficult to draw general conclusions. There is a wealth of cost information contained in the project reports, albeit in different forms, but the benefits proved more difficult to quantify in financial terms.

There are obvious win–win strategies. In particular, those that reduce the heat inputs to a building while also saving energy (and money) at once, such as high-efficiency equipment and lighting, tick both mitigation and adaptation boxes. In practice, designers will know that it is difficult enough to convince clients to adopt sensible low-energy design measures that have quantifiable and relatively short-term benefits. Making a case for strategies which may only pay back in the longer term is significantly more difficult. There are multiple levels of uncertainty facing clients with limited budgets: the uncertainty of future change, of inflation and energy costs, of how the building might be used, and of predicted benefits under conditions of which we have no experience.

Certainly, some clients felt that the longer time periods – up to the 2050s and 2080s – used in the analysis of building performance were too extended. Even the 2020s might appear irrelevant for developers who intend to dispose of an asset shortly after completion. If they do retain the building and lease it out, they would be paying for somewhat uncertain improvements in performance for the benefit not of themselves but of their tenants. Indeed, some of the benefits may not be directly relevant to the occupier of a building but to society as a whole – for example, if the combined effect of measures to reduce cooling demand meant that it was not necessary to increase the capacity of the electricity grid. These wider societal goods cannot be catered for by the market, and regulation would seem to have a role here as the most effective means of serving collective rather than individual interests.

Even though strategies to adapt to climate change may be long term, parts of them could be implemented sequentially. A number of projects highlighted this approach, which would allow measures to be introduced when there is clear evidence that they are required and for decisions to be made in the context of costs at that time.

● ● ●
LEADERSHIP

The government has recognised that it should play a fundamental role in providing leadership in addressing the UK's adaptation response as a whole, and that of the built environment in particular. The intentions outlined by these work streams and deliverables set out in the departmental adaptation plans and the Climate Change Risk Assessment could not be clearer (see opposite). However, it must be recognised that departmental resources are limited and that there may be other pressing issues with more immediate voter appeal than something that might benefit an electorate in the (politically) distant future. Climate change is an insidious process, only perceived when some tipping point is reached that results in a noticeable difference or when extreme events occur, such as the European heatwave of 2003.

The role of extreme events in the development of an adaptation agenda should not be ignored – the case for measures to combat overheating would be easier to make after a series of hot summers. Unfortunately, we may not have the time to wait until the climate makes the case for us. If we do not take the necessary steps in advance of such events, when they do occur we are likely to reach for quick-fix solutions to patch over the consequences rather than to put in place more considered measures with long-term benefits. It is therefore vital that the momentum generated by the Design for Future Climate programme is capitalised on in order to develop a sound, cost-effective, proactive approach to adaptation.

It is encouraging that discussions are already under way between participants in the programme and the Department for Communities and Local Government (DCLG), as noted in case-study examples. However, it should be recognised that the task of embedding adaptation thinking in design practice, standards and, where appropriate, regulation is not trivial. The framework is already in place; now the components must be updated with climate-adapted standards and guidance to provide consistency, setting "reasonable" parameters against which a building design should be tested that are simple enough to be understood and applied reliably by the mainstream, but sufficiently sophisticated to reflect the complex interactions between a building's form, materials and environmental context.

Some areas are already well covered by organisations such as the Environment Agency and CIBSE, which are able to call on relevant expertise and have the authority to develop robust and respected standards and guidance. But there is certainly a role for the government in setting the agenda and allocating responsibility where this is unclear, to ensure that key issues are dealt with systematically.

The planning system was not tested to any great extent by the projects. While it is difficult to imagine that reasonable adaptation measures would cause any additional problems for new buildings, those working on retrofit projects flagged up a real or perceived reluctance by listed-building and planning officers to allow external alterations. It is too early to try to set out adaptation measures that would fall under the category of permitted development, but it would

A DEPARTMENTAL ADAPTATION PLAN

The Departmental Action Plan produced for the Department for Communities and Local Government includes the following "key workstreams" for the built environment:

- **"Communicating the evidence** – including the Climate Change Risk Assessment, with the aim of providing the built-environment sector with latest evidence and usable information on unexpected climate changes and risks, impacts and benefits for the built environment
- **Providing the policy framework** – taking a joined-up approach across government, understanding how different actions connect and identifying options for improvement, including the review of Building Regulations, the development of criteria for measures to be eligible for the Green Deal and the development of National Planning Policy Framework
- **Supporting innovation** – including skills for adaptation in the built environment, knowledge transfer from the Adaptation and Resilience to Climate Change research programme, and take-up of new technology
- **Supporting the delivery landscape** – building collaborative relationships with key organisations, so that industry and professional organisations have sufficient information about the projected changes in climate and are able to support the built-environment sector to make adapted decisions."

Among the "key priorities and deliverables" for its built-environment project are:

- **"Climate data and guidance** – an opportunity to respond to industry concerns that current projections are not easily usable, and to present data in usable formats such as weather files for testing building performance

↳

- **Overheating research** – a priority ... is to better understand the nature of overheating in buildings, the future impact of overheating on the built environment and the case for action. It will investigate what tools and levels of action are needed to assess and prevent overheating, what benefits these might bring and what role, if any, regulation might play alongside work with industry
- **Innovation** – linking to Technology Strategy Board work to support dissemination and learning, building on their successful Design for a Future Climate programme ...
- **Identifying cost-effective adaptation** – contributing to the Adaptation Sub-Committee's work on developing a cost–benefit curve for adaptation."

be useful to review the position once there is sufficient experience and a clearer case for the need for such interventions.

A number of local planning authorities have already set a requirement that climate change should be addressed as part of a planning submission. If this requirement was made more specific – for example, requiring a statement on the steps that have been taken to avoid future overheating – it could present an excellent opportunity to promote serious consideration of the future performance of proposed buildings, but might also present challenges to planning departments as to how to evaluate the statements. One might argue that this level of technical justification would be more appropriately dealt with later in the design process under Building Regulations. However, it is becoming clear that some of our most commonly used building archetypes, such as flats with single-sided ventilation, may need to be re-examined to ensure that they are still valid in the context of future climate. The use of such archetypes is fundamental in establishing the density and development potential of sites and it is essential that the implications of adopting them are examined early enough to avoid the presumption of unrealistic land values and developmental blind alleys.

Of course, all buildings are existing buildings once built, and they may need to go through a planning-approval process in order to implement a sequential adaptation strategy as interventions become necessary. Where a new building depends on visible alterations to adapt to a future climate, should these measures be shown in the original application in order to establish long-term rights to make them? Conversely, should a building where the quality of the architecture relies on shading, for example, be required to include these when first built or can it be built "bald", ready to accept the measures when they become necessary?

A more pressing issue is that measures undertaken in order to meet the mitigation agenda, such as the thermal upgrading of homes under the Green Deal, may have unintended consequences. For example, once they are insulated and sealed up, these homes may be more susceptible to summer overheating under current conditions – let alone in the future. It would be sensible to include a requirement to demonstrate that overheating has been considered as part of any Building Regulations approvals process, and to issue guidance to cover alterations that do not require approval. The TSB-funded Retrofit for the Future programme of projects to upgrade and monitor existing homes should provide useful data on this, and perhaps the monitoring regime should be continued and adapted to provide information more relevant to overheating, in addition to energy performance.

Some teams cited regulatory standards from warmer locations in other countries as a potential basis for criteria for our future climate, and it would be sensible to research these systematically to pick out best practice – particularly in dealing with regional variation when considering overheating.

The combined experience of the teams involved in the Design for Future Climate projects is a rich resource on which to draw to develop necessary consensus, based on practical and detailed investigation of the issues. Projects in the second tranche are now starting to reach their conclusions and it will be interesting to see how their approaches have developed, particularly among those who have been able to build on their experience from the first tranche.

Adaptation to a changing climate is a challenge that modern civilisation has never faced before, and it brings huge opportunities for designers. Those involved in the Design for Future Climate projects have amply demonstrated that design teams have the intelligence, technical skill, design flair and enthusiasm to rise to the challenge of adaptation and seize the opportunities that it presents.

Appendix 1

DETAILS OF THE DESIGN FOR FUTURE CLIMATE PROJECTS

DFCC PROJECT NUMBER	NAME	PROJECT TEAM *Client in italics* **Project lead in bold**	BUILDING TYPE	PROJECT DESCRIPTION ■ CIBSE Case Study available
1	London School of Hygiene and Tropical Medicine	*London School of Hygiene and Tropical Medicine* **AECOM** Day England Stevenson Marsh Davis Langdon AECOM	Laboratory	■ Grade II-listed education building refurbishment, including specialist, highly serviced labs and general office and teaching spaces
2	County Hall, Trowbridge Wiltshire	*Wiltshire County Council* **WSP Built Ecology** Stride Treglown	Offices	■ Refurbishment of 1930s stone-construction four-storey listed building, with a 1970s concrete-framed extension
3	Extra Care 4 Exeter	*Exeter City Council* **Gale & Snowdon** University of Exeter Jenkins Hansford Partnership	Care home	■ New-build care home on a brownfield site with 50 self-contained dwellings for elderly people and those with moderate dementia
4	Ellingham Primary School, Kingston, London	*Royal Borough of Kingston* **ECD Architects** Environmental Design Associates Silcock Dawson & Partners Centre for Alternative Technology Keegans Clark Smith Partnership	School	Demolition and rebuilding of a primary school
5	Great Ormond Street Hospital, Phase 2B	*Great Ormond Street Hospital* **WSP Built Ecology** Llewelyn Davies Yeang Gardiner & Theobald University College London	Hospital	■ New cardiac wing for children's hospital, converting existing concrete-framed building to steel-framed seven-storey wing
6	University of Greenwich School of Architecture, Design & Construction	*University of Greenwich* **Hoare Lea** Heneghan Peng Architects University of Manchester Fanshawe Alan Baxter	University	■ New building for University of Greenwich construction and architecture school, including a learning-resource centre
7	Oxford University Press Offices, D-Wing Extension	*Oxford University Press* **Hoare Lea** Berman Guedes Stretton University of Manchester Baqus Sworn King Price & Myers	Office	Part demolition, part refurbishment of existing office building, including construction of new atrium space and three-storey office building with basement
8	University of Sheffield Engineering Graduate School	*University of Sheffield Graduate School* **Arup** Bond Bryan Architects Turner & Townsend	University	New university teaching space, including large lecture theatres, general teaching spaces, labs, facilities for postgraduate and doctorate work space, and offices
9	British Trimmings extra-care scheme, Leek, Staffordshire	*Harvest Housing Group* **Triangle Architects** Leeds School of Architecture The Energy Council S I Sealy & Associates SDA Consulting ABA Consulting	Care home	New-build development with 87 self-contained flats for elderly people with extra-care needs
10	North West Cambridge NWC	*University of Cambridge* **AECOM**	Housing	Masterplan for University of Cambridge land, including analysis of Urban Heat Island effect and summer overheating

DFCC PROJECT NUMBER	NAME	PROJECT TEAM *Client in italics* **Project lead in bold**	BUILDING TYPE	PROJECT DESCRIPTION ■ CIBSE Case Study available
11	Wyre Forest primary schools, Worcestershire	***Worcestershire County Council*** Sjölander da Cruz Max Fordham University of Manchester Grant Associates Bridgewater & Coulton Shire Consulting Robert Bray Associates	Schools	Three primary schools, including two new builds and one existing building
12	Church View, Doncaster	*Doncaster Development Community Trust* **Bauman Lyons Architects** Arup Herriot Watt University Oxford Brookes University Estell Warren Latz + Partners Bovis Lend Lease Creative Space Management	Office	■ Refurbishment of an art college building in a conservation area to become multi-occupancy serviced offices
13	Harnessing nanotechnology to combat climate change, Central Saint Martins, University of the Arts London	*Central Saint Martins College, University of the Arts* **Stanton Williams** Atelier Ten Nanoforce Technology	University	Redevelopment of Grade II-listed granary building into a new university campus, as part of King's Cross Central redevelopment
14	Climate Adaptation Plan, Marks & Spencer	*Marks & Spencer* **Deloitte** Darton EGS Troup Bywaters & Anders Walker Institute Gleeds WSP	Retail	Developing a climate-adaptation "tool kit" for the refurbishment of retail units in the UK, typical of 97% of Marks & Spencer's projects
15	PortZED, Brighton and Hove	***BohoGreen*** APZED (Alan Philips Architects + ZEDFactory) Bobby Gilbert & Associates Hemsley Orrell Partnership Monson Acoustic Associates Hayes Mackenzie	Housing	Mixed-use seafront development
16	100 City Road, London	*Derwent London* **Arup** Allford Hall Monaghan Morris Adams Kara Taylor II Davis Langdon AECOM	Office	■ Speculative office development
17	Harris Academy, Purley	*Harris Federation and London Borough of Croydon* **Aedas** VZDV Consulting Engineers Wilmott Dixon	School	■ 4,000m² part-new-build, part-refurbishment school for 11–18 year olds
18	Welland Primary School, Peterborough	*Peterborough City Council Children's Services* **AECOM** Kier Eastern Woods Hardwick Mott MacDonald ACD Landscape Architects Davis Langdon AECOM	School	New single-storey school with natural ventilation

DFCC PROJECT NUMBER	NAME	PROJECT TEAM *Client in italics* **Project lead in bold**	BUILDING TYPE	PROJECT DESCRIPTION ■ CIBSE Case Study available
19	11–16 Phase School, Ebbw Vale	*Blaenau Gwent County Borough Council* **Building Design Partnership** Davis Langdon AECOM Willmott Dixon	School	New 11–16 school on the Ebbw Vale Welsh "EcoTown" site
20	Technical Hub @ EBI, Hinxton	*The Wellcome Trust, European Bioinformatics Institute* **AECOM** Abell Knepp Turner & Townsend	Laboratories, offices	New research and laboratory testing space for the Wellcome Trust Genome Campus in South Cambridgeshire
21	The Mill, Cardiff	*Welsh Assembly Government* **White Design** Forum for the Future Faithful + Gould Savills	Housing	New residential quarter of 900 homes in central Cardiff on a linear site bounded by railway lines
22	The Royal Academy for Deaf Education, Exeter	*The Royal Academy for Deaf Education, Exeter* **Skelly & Couch** dRMM Architects The Royal Botanic Gardens, Kew University of Kent Ansys UK	School	New school building with an ETFE-and-timber domed roof
23	New Admiral Insurance Headquarters, Cardiff	*Admiral Insurance/Stoford* **BRE** Glenn Howells Architects Hoare Lea	Office	90-year adaptation plan for new 1,860m² FTSE100 headquarters
24	NW Bicester Eco Development	*P3 Eco (Bicester) and A2 Dominion Group* **Hyder Consulting** Farrells Low Carbon Building Research Group, Oxford Brookes University	Housing	Masterplan for EcoTown, first phase – comprising 350 homes, primary school, care home, business innovation centre and cooperative grocery store
25	Edge Lane: TIME project, Liverpool	*Liverpool and Sefton Health Partnership* **Medical Architecture and Art Projects** Low Carbon Building Research Group, Oxford Brookes University, Mott MacDonald Fulcrum Tony Danford Arup Camlin Lonsdale Davis Langdon AECOM	Hospital/ care home	6,074m² new in-patient mental-health facility with 75 beds, on brownfield contaminated site
26	Cornwall Council Office Rationalisation Programme	***Cornwall Council Property Services*** HLM Architects Bradbury Wynter Cole Architects Hoare Lea Parsons Brinkerhoff University of Exeter Cyril Sweett	Offices	Reducing Cornwall Council's office premises from 78 to 30; refurbishment of three existing offices

Appendix 2

CLIMATE VARIABLES INCLUDED IN THE UKCP09 CLIMATE PROJECTIONS

More information on the UKCP09 climate projections is available at:

http://ukclimateprojections.defra.gov.uk

Variables over land areas

1. Mean temperature
2. Mean daily maximum temperature
3. Mean daily minimum temperature
4. 99th percentile of daily maximum temperature in a season (warmest day of the season)
5. 1st percentile of daily maximum temperature in a season (coolest day of the season)
6. 99th percentile of the daily minimum temperature in a season (warmest night of the season)
7. 1st percentile of daily minimum temperature in a season (coldest day of the season)
8. Precipitation rate
9. 99th percentile of daily precipitation rate in the season (wettest day of the season)
10. Specific humidity
11. Relative humidity
12. Total cloud
13. Net surface long-wave flux
14. Net surface short-wave flux
15. Total downward short-wave flux
16. Mean sea-level pressure

Variables over marine regions

17. Mean air temperature
18. Precipitation rate
19. Total cloud
20. Mean sea-level pressure

UK CLIMATE PROJECTION MAPS

For the original Design for Future Climate report, the UK Climate Impacts Programme (UKCIP) developed a series of maps based on the UKCP09 projections, to provide building designers with a broad understanding of the changes they needed to consider.

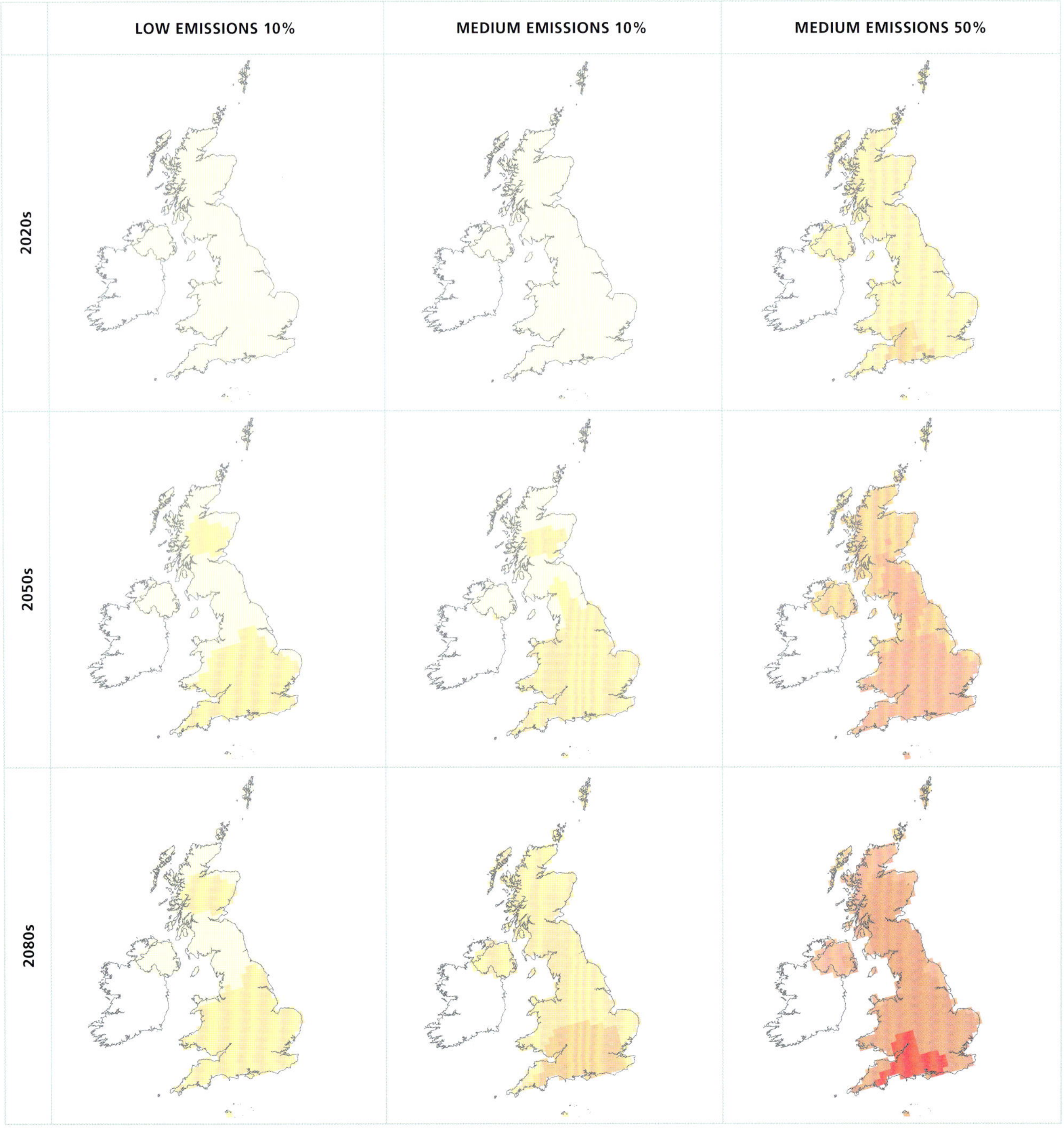

	LOW EMISSIONS 10%	MEDIUM EMISSIONS 10%	MEDIUM EMISSIONS 50%
2020s			
2050s			
2080s			

This map is one of a series produced by UKCIP based on the UKCP09 climate projections. It shows projected changes in maximum summer temperatures in °C, relative to the 1961–1990 baseline, for a range of scenarios and percentiles.

Source: Data © UKCP09, 2009. Interpretation © UKCIP, 2010.

The full set can be viewed in the original report on the Technology Strategy Board website, at www.innovate.org/adaptation. An example is provided here, showing projected changes in maximum summer temperatures in °C for a range of scenarios and percentiles. See Chapter 2 for an explanation.

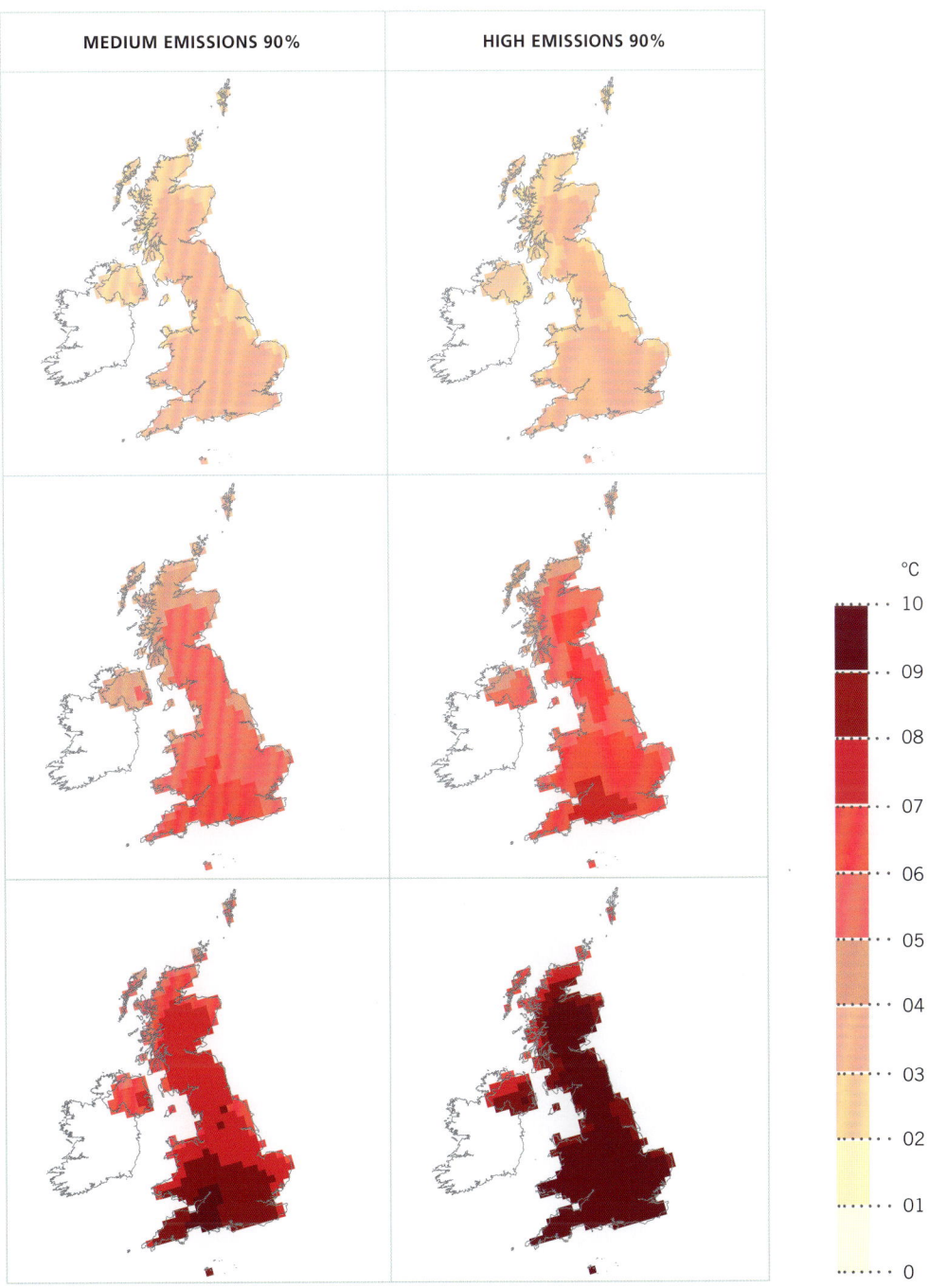

Notes

1   https://connect.innovateuk.org

2   Parker, D.E., Legg, T.P. and Folland C.K. (1992) 'A new daily Central England Temperature Series, 1772–1991', International Journal of Climatology, Vol 12, pp317–42. Available at: http://www.metoffice.gov.uk/hadobs/hadcet/Parker_etalIJOC1992_dailyCET.pdf

3   TM48: The Use of Climate Change Scenarios for Building Simulation: the CIBSE Future Weather Years, CIBSE, 2009.

4   Burberry, P. (1970) 'Environment and Services', Longman.

5   Table 1.4, Guide A: Environmental design, 7th Edition, Issue 2, CIBSE, 2007.

6   Ibid., Table 1.2.

7   Humphreys, M.A., Rijal, H.B. and Nicol, J.F. (2010) 'Examining and developing the adaptive relation between climate and thermal comfort indoors.' Proceedings of Conference: Adapting to Change: New Thinking on Comfort, Cumberland Lodge, Windsor, UK, Available at: www.nceub.commoncense.info/uploads/22-01-05-Humphreys.pdf

8   Gareth Roberts (Sturgis Carbon Profiling), 'Whole Life Carbon: Prestige Offices', Building magazine, May 18 2012, pp 52–56.

9   Coley, D. (2011) 'A first look at the effect of outdoor planting on the indoor environment of St Loyes Residence', University of Exeter.

10 Archer, S. (2011) 'Green walls: Thermal and Hydrological Costs and Benefits, E-Futures Mini Project.' Available at: http://e-futures.group.shef.ac.uk/publications/pdf/118_3.%20 Stuart%20Archer%20Summary.pdf

11 Pochee, H., Dawson, G., Burgon, P. and Bentham, T. (2012) 'An Analysis of the Benefits and Drawbacks of Exposed Thermal Mass in Modern, Well Insulated Buildings', Max Fordham.

12 Auliciems, A. and Szokolay, S.V. (1997, revised 2007) 'Thermal Comfort, PLEA Notes: Passive and Low Energy Architecture International', University of Queensland.

13 Ibid.

14 'Climate Change and Innovation in House Building – Designing Out Risk', NHBC Foundation, 2007.

15 'The Vulnerability of UK Property to Windstorm Damage', Association of British Insurers, 2003.

16 WUFI (Warme Und Feuchte Instationar) software, produced by the Fraunhofer Institute to calculate the coupled heat and moisture transfer in building fabric.
Available at: www.wufi-pro.com

17 'Ecological Site Classification to assist in tree-species choice, based on suitability to current and future climates (to 2050).' Available at: www.forestry.gov.uk

18 Sanderson, M. (2010) 'Changes in the frequency of extreme rainfall events for selected towns and Cities', Met Office for Ofwat.

19 Flather, M. (2010) 'The effects of field sizes on rural flood hydrographs in rural catchments.' Unpublished MSc thesis, University College London.

20 Rodda, H.J.E., Berger, A. and Muir-Wood, R. (2002) 'A stochastic model for exploring extreme flood events in the UK.' Proceedings of The Extremes of the Extremes: Extraordinary Flood Symposium, Reykjavik, Iceland.

21 https://connect.innovateuk.org

Index

Note: page numbers in italics refer to figures;
page numbers in bold refer to tables